NORTH CAROLINA
STATE BOARD OF COMMUNITY COLLEGES
LIBRARIES
ASHEVILLE-BUNCOMBE TECHNICAL COMMUNITY COLLEGE

DISCARDED

APR 15 2025

READINGS IN HAZARD CONTROL AND HAZARDOUS MATERIALS

Consulting Editor

David V. MacCollum, CSP, P.E.
Sierra Vista, Arizona

American Society of Safety Engineers
850 Busse Highway
Park Ridge, Illinois 60068
312/692-4121

ISBN No. 0-939874-70-9

Copyright © 1985

FOREWORD

A clear understanding as to what constitutes a hazard is needed when addressing the control of hazards or hazardous and toxic materials. We all know that a hazard is an unsafe physical condition, but we often overlook the fact that a hazard exists in one of three modes:

1. LATENT: A hazard in an unused mode may not be readily perceived by an employer or user while existing in the latent form on the drawing board or as a proposed material to be used in a process.

2. ARMED: The hazard is in a used or operational mode and exposed to circumstances in which harm can occur. However, the situation may still not be perceived to be dangerous by the user or employee.

3. ACTIVE: The hazard is now in the mode where it can cause harm, and unfortunately it is still not recognized to allow for escape or for initiation of appropriate safeguards to avoid or prevent injury.

From studies of human behavior, we realize that it is foreseeable that people may unintentionally trigger hazards into active modes. Accordingly, the active mode of a hazard must be anticipated by the safety engineer or safety professional and not left unabated for the entrapment of the unsuspecting. The identified hazard now needs to be abated or controlled by applying a system safety order of precedence.

1. DESIGN TO ELIMINATE OR MINIMIZE THE HAZARD. The major effort throughout the design phase must be to select appropriate safety design features to eliminate the hazard, render it "fail-safe," or provide redundancy through use of back-up components to avoid a failure mode that will cause the hazard to become active.

2. GUARD THE HAZARD. Hazards that cannot be totally eliminated through design selection in Precedent 1 above must be reduced to an acceptable level of risk through the use of appropriate safety devices that will guard, isolate or otherwise render the hazard effectively unarmed or inert. The hazard should be either made inaccessible or nearby personnel should be protected from the hazard. For example, management should guard moving fan blades, provide rollover protection (ROPS) and make use of interlocks that prevent access to the interior of a high-voltage cabinet by automatically de-energizing the power source when the cabinet door is opened.

3. WARNING DEVICES. Where it is not possible to eliminate totally the existence of a hazard or to provide the aforementioned guards, devices should then be employed for the timely detection of the hazard and for the generation of an active warning signal. Audible and/or flashing warning signals and their application must be designed to allow people to avoid the hazard. Moreover, these systems must be standardized within like types of systems. Passive warnings (such as signs or labels) must be explicit, stating just what the hazard is, its resulting harm and how to avoid it. A passive warning should not be used as a substitute for

Precedents 1 or 2. Rather, a passive warning is best used to acquaint the user as to the need for the safeguard identified and used in accordance with Precedents 1 and 2.

4. SPECIAL PROCEDURES. Where it is not possible to eliminate or reduce the magnitude of a hazard through design or with guards, isolation or the use of safety and warning devices, then management must assure special operating procedures, training and audits to guarantee that a viable, continuing, regimen is established and maintained to avoid the hazard.

To achieve an acceptable level of hazard reduction, a complete safety network of physical safeguards must be established for all high-risk, critical hazards. Often to achieve this goal, it is necessary for one or more controls from each order of precedence to be employed. The effectiveness of a specific hazard control device is its ability to serve as part of a "team" and it should not be discarded simply because it cannot be totally reliable for all circumstances. To reliably assure the life and safety of others, redundancy must be created through use of a variety of safety design features that minimize the hazard, provide physical barriers, give warning or require special procedures. All will collectively prevent the hazard from being armed or activated.

Hazards have three main avenues of entry into the workplace:

1. THROUGH UNSAFE MATERIALS-HANDLING OR PRODUCTION EQUIPMENT, OTHER MACHINERY OR POWER TOOLS. The safety director needs to screen purchases to assure for inclusion of available safety features or to see if those provided are proper for the intended use.

2. THROUGH UNSAFE PRODUCTION METHODOLOGY OR SERVICE PROCEDURES. The safety director must be the principal advisor to the chief executive officer on all matters relating to safety. He must be privy to all existing and contemplated activities and must have authority to assure for remedial action.

3. THROUGH WORK DONE BY NEGLIGENT CONTRACTORS. In many instances, the function of safety in building a new facility is inappropriately transferred to the designer and the contractor. Landowners have a non-delegable duty since the landowners' personnel may be exposed to the contractors' hazardous activities or the contractors' personnel may be exposed to the landowners' hazardous activities which can be controlled.

    A. Evaluate the design of the proposed project to eliminate hazards arising from:

        a. Operation or use upon completion.

        b. Construction methods and practices.

B. Evaluate special safety features and conditions in the design and specifications arising from a and b above.

C. Arrange for a pre-bid safety conference to assure that the prospective bidders are acquainted with design and construction safety requirements.

D. Evaluate the safety program of all prospective bidders, their proposed construction methods, and equipment proposals to eliminate contractors who would be unable to complete the project safely for lack of correct equipment, an overall effective safety program, and/or because of previous unsafe performance.

E. Hold a pre-notice to proceed conference with the successful bidder and record and define in the minutes the following:

   a. Chain of command and scope of safety authority.

   b. Required safeguards.

   c. Methods and procedures for controlling hazardous equipment, materials, or activities.

   d. Accountability, by name and position of manager or supervisor directly responsible for specific hazardous activities or operations which require specific safeguards.

F. Hold monthly management-level safety conferences to involve the responsible landowner's representative, the prime contractor's general manager, and any sub-contractors' superintendents to address progress for the control of all hazardous conditions and for planning control of anticipated hazards. It is at such meetings that the safety director for the landowner can assure that each hazard is identified and controls adequate. Minute-taking at such meetings will assure for accountability of supervisory personnel assigned to oversee the hazardous activities and the use of specific safeguards.

Unabated hazards or hazardous materials create an inherently dangerous workplace or procedure. In terms of safety engineering, if any foreseeable or anticipated exposure or choice of use can result in serious harm to the user or others, then that exposure or choice of use is inherently and intrinsically dangerous. Control measures are necessary as defined in the system safety order of precedence to assure for the abatement or control of all hazards.

Hazard control is a great deal more sophisticated than the "square one" approach of assuming that all employees will reliably identify each and every hazard and then exercise perfect judgment and performance 100 percent of the time for all circumstances. Assuming that safe work practices will overcome a physical hazard or unsafe condition is an invitation to disaster. Human factor principles clearly define machinery, facilities or products as hazardous when they are error-provocative, violate user expectations, require performance beyond what the user can deliver, induce fatigue or create an unpleasant or undesirable circumstance.

Effective hazard control recognizes the variables of human behavior and dictates that safe systems be designed to take into account normal, foreseeable, human performance.

Physical hazards as sources of misdirected energy or as threats to health can be categorized into easily identifiable groups. The articles from Professional Safety carried in this publication fall within four basic groups: Environment, Mechanical, Chemical Reaction and Electrical. Four other articles appear under the "General" category.

When the knowledge of a competent safety professional is used in combination with the absolute support of top management to control all hazards, effective hazard control is achievable.

David V. MacCollum, CSP, P.E.
Consulting Editor

David V. MacCollum, CSP, P.E., is president of the safety consulting firm of David V. MacCollum, Ltd. He has been in private practice for over 13 years. He is a Past President of the American Society of Safety Engineers (1975-76) and twice the recipient of the American Society of Safety Engineers/ Veterans of Safety First Place Technical Paper Award in 1969 and 1984.

His safety expertise has interfaced with membership in many professional groups: a senior member of System Safety Society; and a member of the Human Factors Society, National Society of Professional Engineers, Mining Engineers, and the National Safety Council.

He testified before the U.S. Senate's March, 1970 hearing for the Product Safety Commission on the hazards of unvented heaters and also before its April, 1977 hearing on Product Liability Insurance.

He served on the advisory body which drafted the Arizona Occupational Safety Act and was a member of the Arizona Review Commission of Appeals for state citations.

In 1972-73 he was retained as an instructor at the University of Arizona for a series of courses on Systems Safety, Safety Management, and Safety Program Evaluation, and developed a graduate safety management course for the University of Arizona and also other special safety engineering programs for it and Arizona State University, Michigan Technological Institute, University of Oklahoma, University of Wisconsin and the National Institute of Occupational Safety & Health (NIOSH).

He served on the U.S. Department of Labor's Construction Safety Advisory Committee, 1969-1972; was Chairman of the sub-committee for Subpart V dealing with power transmission and distribution; and was a member of the board investigating tunnel disasters.

He developed design citeria for rollover protection (ROPS) in 1958 which was adopted by the U.S. Army Corps of Engineers, U.S. Bureau of Reclamation, and State of Oregon, 10 years prior to the development of rollover standards by the Society of Automotive Engineers (SAE) and American Society of Agricultural Engineers (ASAE).

His previous career was with U.S. Department of Army from 1955-1972 as Director of Safety for the Strategic Communications Command at Fort Huachuca, Arizona, a world-wide command with 16 subcommands.

He was also Safety Director for the Electric Proving Ground where he developed system safety doctrine for safety evaluation of a broad range of army combat and support equipment.

He was Safety Director for two combat infantry divisions and an amphibious combat engineer battalion. He started his Federal service as Safety Engineer for the Portland District, Corps of Engineers, where he developed safety standards for reverse signal alarms, ammonia nitrate explosives, anti-two-blocking devices for cranes, crane testing, etc., for reducing hazards on civil works construction.

His career commenced in 1951 as a Safety Engineer for the State of Oregon Industrial Accident Commission. In his private life, he has served his community as a director of an electric utility for nine years and served on several commissions for the city of Sierra Vista.

Over some 30 years his safety expertise has been international in nature. He has presented papers to such groups as British Ministry of Technology, American Medical Association, and other professional engineering societies in addition to ASSE and the NSC. His private practice has extended to overseas manufacturers, U.S. trade associations, and as an expert witness in a broad variety of hazard-related litigation.

READINGS IN HAZARD CONTROL AND HAZARDOUS MATERIALS

Table of Contents

GENERAL

HOW DO YOU KNOW YOUR HAZARD CONTROL PROGRAM IS EFFECTIVE? 1
Fred A. Manuele (June, 1981)

ENGINEERING PROJECT PLANNER, A WAY
TO ENGINEER OUT UNSAFE CONDITIONS 8
Donald J. Eckenfelder and Charles E. Zaledonis (November, 1976)

WARNING LABEL DESIGN 14
Michael W. Riley, David J. Cochran and John E. Deacy (October, 1981)

NEW DIMENSIONS IN THE TORTIOUS FAILURE TO WARN 17
Harry M. Philo (November, 1982)

ENVIRONMENT

FALL ACCIDENT PATTERNS 24
H. Harvey Cohen and D. M. J. Compton (June, 1982)

ENTERING AND EXITING ELEVATED VEHICLES 31
Ron Hurst and Tarek Khalil (September, 1984)

CONFINED SPACE ENTRY 38
Steve P. Krivan (September, 1982)

MECHANICAL

CRITICAL HAZARD ANALYSIS OF CRANE DESIGN 43
David V. MacCollum (January, 1980)

MACHINE SAFEGUARDING AND SAFETY MANAGEMENT 49
D. A. Colling (August, 1982)

EXPOSURE REDUCTION IN LP-GAS INSTALLATIONS 53
Hugh F. Keepers (August, 1979)

LESSONS FROM 25 YEARS OF ROPS 57
David V. MacCollum (January, 1984)

CHEMICAL REACTION

HAZARD CONTROL OF LIQUID OXYGEN SYSTEMS 64
William W. Allison (January, 1979)

HAZARD CONTROL OF OXYGEN SYSTEMS: AN ADDENDA 69
William W. Allison (October, 1979)

CONTROLLING TOXIC SUBSTANCES AND HAZARDOUS MATERIALS 73
George B. Stanton (June, 1980)

TRANSPORTING, LOADING, AND UNLOADING OF HAZARDOUS MATERIALS 80
William S. Wood (July, 1976)

AN EASY WAY TO DECIDE ABOUT HAZARDOUS LIQUIDS 86
Azmi P. Imad and Cheri L. Watson (April, 1980)

PESTICIDE SAFETY AND HEALTH 91
C. Saran, Linda L. Sims, Paul M. Hughes
and Stephen A. Adam (February, 1983)

HEALTH HAZARDS IN THE WORKPLACE 96
Ernest Mastromatteo (February, 1982)

TRAINING FOR TOXIC MATERIALS HANDLING.............................. 99
James E. Gillan (April, 1978)

TOXIC EFFECTS OF WOOD DUST EXPOSURE 102
Angelo Meola (March, 1984)

ELECTRICAL

THE CLAMPDOWN ON ELECTRICAL HAZARDS................................ 106
Milton Leonard (September, 1975)

EXPLOITING TECHNOLOGY FOR ELECTRICAL SAFETY 111
Alan Krigman (April, 1975)

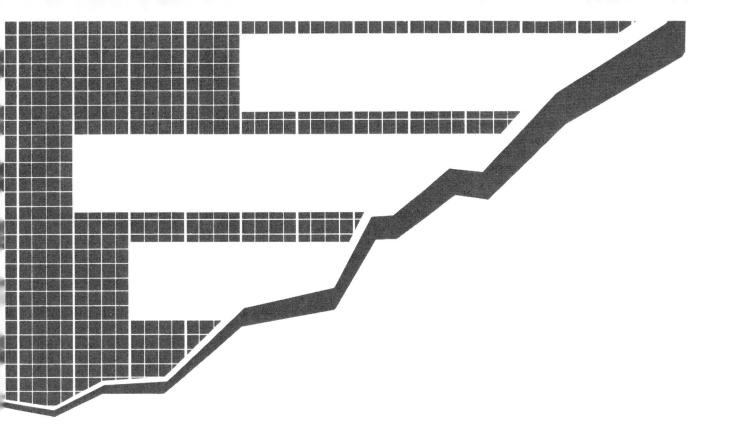

How do you know your hazard control program is effective?

by Fred A. Manuele

A review of the literature on the measurement of hazard control performance requires the conclusion that precisely accurate evaluation systems have not been established, that difficulty arises in applying the requirements of valid measurement concepts to hazard control and that the best approach is to use several measurement systems to evaluate the quality of hazard control management.

Hazard control practitioners who are truly professionals must seek and promote the development of measurement systems through which the effectiveness of the actions they propose can be measured. Performance examination is a necessity, and we ought to be doing the best we can, recognizing and accepting that any measurement systems adopted have their shortcomings.

That combination of measurement systems which I propose you consider includes OSHA incident rates, worker's compensation costs, The Critical Incident Technique and evaluations of the effectiveness of hazard control programs.

None of these systems will meet all of the requirements of accurate measurement concepts. A composite listing follows of the characteristics of a good measuring technique, developed from the writings of Firenze in "The Process of Hazard Control," Bird and Loftus in "Loss Control Management," Tarrants in "Applying Measurement Concepts to the Appraisal of Safety Performance," and Jacobs in "Towards More Effective Safety Measurement Systems." A measurement system should be:

1. Administratively feasible
2. Adaptable to the range of characteristics to be evaluated
3. Constant
4. Quantifiable
5. Sensitive to change
6. Valid in relation to what it is supposed to represent

7. Capable of duplication with the same results from the same items measured
8. Objective, efficient and free from error

As Tarrants wrote, "We have identified what we would like to have if we could wave our magic wand and create a perfect measurement instrument. These items, taken collectively, represent the star toward which we should direct our attempts to improve our safety measures." Recognizing the theoretical ideal, and that we have no magic wand to wave, I would like to explore the benefits of the systems I propose be considered.

Incident rates

OSHA incident rates are, surely, historical. They reflect what has happened. Arguments against the use of historical data as a measure of safety performance suggest that the data is not sufficiently sensitive, that all incidents are not reported, that the data does not necessarily reflect the current status of safety performance and that historical information cannot logically be used for prediction purposes.

All of those reasons had greater validity when injury and illness data was computed in accordance with the American National Standards Institute "Method of Recording and Measuring Work Injury Experience" Z16.1, since only those injuries with relatively high severity were recorded. They are less valid as respects data recorded under OSHA since the definition of recordable injuries and illnesses often requires as many as 3 to 4 times more recordings than under the ANSI method, thus achieving a greater sensitivity to what actually occurs in the work place.

Of course, the thoroughness of reporting has a bearing on the credibility of the data.

While recordable incident rates do not absolutely reflect the current status of safety performance, they provide one usable measure. Cautions are expressed in the literature on the use of historical injury data when the work force is small. Obviously, the credibility of the data increases with the population. At 500,000 hours worked, OSHA data should have considerable reliability, except for severity of injury potential.

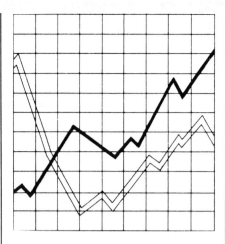

Severe injuries aren't all that frequent. Computing a rate of lost work days provides a partial indicator of severity, but even that may not be adequate as a measure of potential. Extending the concept of The Critical Incident Technique particularly to include observations of severity potential would provide another measure of value.

It's also argued that historical OSHA incident rates don't relate to the present or the future, since people, machines and the environment are constantly changing. My counter-argument would be that the level of safety in an establishment is achieved over a period of considerable time. Except in the unusual circumstance, the level doesn't change quickly, for better or worse. In either direction, change takes time.

When comparing OSHA incident rates to those of a similar operation, it's prudent to get as close to an apples to apples relationship as possible, which isn't always easy. Within the same organization, operations producing the same product have their differences. We often find ourselves recommending that a location compare its record against its own past performance and set reasonable improvement goals for the future.

Considering other measures as well, you would know that your hazard control program is effective if you achieved your OSHA incident rate goals.

Severity of injury potential requires its own consideration. An Italian economist, Vilfredo Pareto, observed in the 1890's that "The significant items in a given group normally constitute a relatively small portion of the total." Through a study of reported injuries made by Employers of Wausau, it was determined that:

—67% of injuries account for only 6% of total injury cost
—33% of injuries account for 94% of injury cost
—2% of injuries account for 50% of injury cost

Cost measurement systems

Consideration should be given to cost measurement systems for one very good reason. Those decision-makers whom hazard control practitioners should be influencing understand the language of money. They usually don't understand other measures as well.

Particularly within the past few years, operating executives have had to give a much greater attention to worker's compensation costs and, thereby, the quality of their hazard control programs. It is not unusual for worker's compensation costs to have doubled in the past 3 or 4 years, while OSHA incident rates may have remained pretty much the same. Increases in worker's compensation costs don't necessarily mean that the quality of hazard control programs has deteriorated. Ignoring the use of costs deprives the hazard control practitioner of a measure that could influence decision makers toward achieving more effective programs.

Most states, in response to the recommendations of the National Commission on State Workmen's Compensation Laws, have greatly increased benefits available and extended the coverage of benefits. Taking into consideration only the maximum weekly payments available for permanent total disabilities and temporary total disabilities, during the period from January of 1976 to January of 1979, five states increased available benefits over 90%, in eight states, the increases were 60%-90%, and in nine states the increases ranged between 30%-60%. In the remaining states, the increases were under 30%.

Thus, because of increases in benefit levels, performance measures relating one year to another require explanation. Nevertheless, measurement of actual injury and illness costs have a bearing on the significance of hazard control in management decisions which establish priorities and allocate resources, and attempts should be made to evaluate cost trends.

Considering other measures as well, you would know that your hazard control program was effective if cost goals were established, and met.

Both OSHA incident rates and worker's compensation cost data are historical measures. While several writers have set forth the shortcomings in the use of historical data as performance measures, it was of interest that Daniel C. Petersen in his book titled "Techniques of Safety Management" stated that "The past performance of any group is the best standard to use as a guide for present performance."

Critical Incident Technique

Several texts and published papers contain reviews of the Critical Incident Technique. In "Safety Management" by Grimaldi and Simonds, this footnote is given as respects the origin of the procedure:

> The Critical Incident Technique originated in World War II as a result of studies to determine the likely causes of military aircraft accidents. In the method, an observer interviews a sampling of persons listing their recall of unsafe acts and conditions occurring in their work. The stratifications are selected according to degree of hazard, type of exposure, degree of exposure, and other factors considered significant to the investigation. The purpose is to identify the critically unsafe acts and conditions in the operation and correct them before they occasion accidents.

Allow me to emphasize that the purpose of The Critical Incident Technique is to "identify the *critically* unsafe acts and conditions in the operation and correct them before they occasion accidents."

The Critical Incident Technique allows both a measure of safety performance and an identification of practices or conditions that need correction. If, in the application of The Technique, an emphasis was given to severity of injury potential, it would have greater value since severe injuries are relatively rare and don't show up very often in OSHA statistics or in accident investigation procedures.

To apply The Critical Incident Technique, a hazard control practitioner would establish an appropriate sample of the work population in

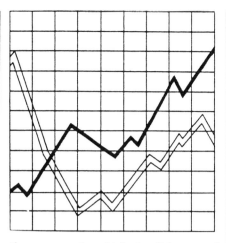

those areas in which the "degree of hazard" had been established as warranting attention. This method is suggested, as a beginning, to achieve the greatest benefit in relation to the expense.

Hazard control practitioners would question persons in the sample who have performed tasks within the environments selected, asking that they recall and describe unsafe practices they have performed, or unsafe conditions with which they are familiar. Quite probably, the selected employees would be much more willing to talk about unsafe practices and conditions which were not the causes of accidents, particularly involving themselves, than those that were.

A skill is necessary in the interview process. Participants would have to be encouraged to be forthright in their comments so that as many probable causes as possible could be identified and classified as respects "degree of hazard."

Corrective action in relation to identified unsafe acts and conditions would follow, and should be at two levels. To begin with, action should be proposed to correct those specific unsafe acts and conditions identified. In addition, inquiry would profitably be made as respects those management systems that should have prevented the occurrence or existence of the identified unsafe acts and conditions, to determine how those systems can be improved.

The greatest benefit derived from application of The Critical Incident Technique could very well be the improvements which result in hazard control program management. In addition, a measurement of the quality of safe performance can be obtained, as well as the correction of the individual unsafe practices and conditions noted.

This procedure is not a one-time thing. It needs application and reapplication to be effective, in relation to the constant changes that are typical in the person-machine-environmental situations that commonly exist.

Dr. Tarrants has been an outspoken advocate of The Critical Incident Technique. In his paper titled "Applying Measurement Concepts to the Appraisal of Safety Performance," he mentioned an extensive study made to evaluate the usefulness of the Technique. His conclusions were:

- The Critical Incident Technique dependably reveals causal factors in terms of errors and unsafe conditions which lead to industrial accidents
- The Technique is able to identify causal factors associated with both injurious and non-injurious accidents
- The Technique reveals a greater amount of information about accident causes than presently available methods of accident study and provides a more sensitive measure of total accident performance
- The causes of non-injurious accidents as identified by The Critical Incident Technique can be used to identify sources of potentially injurious accidents
- Use of The Critical Incident Technique to identify accident causes is feasible

This system of safety performance measurement has considerable value because of its anticipatory characteristic: it attempts to identify potential problems before they occur. An emphasis on high severity potential would result in a greater impact.

Considering other measures as well, you would know that your hazard control program was effective, if application of The Critical Incident Technique established that the unsafe practices and conditions identified were within what you considered to be tolerable limits. Until the procedure is applied, a suitable performance level cannot be established. There are no published norms.

Importance of audits

Because of an extensive involvement of the staff at M&M Protection

Consultants, we are intrigued with the benefits which derive from audits of the effectiveness of hazard control programs as performance measures. Certain premises were established in the development of our procedure:

- Management is the planning, organization, direction and control of activities to achieve desired goals
- It is necessary in a successful business process to set policy, establish procedure, assign responsibility, institute an accountability system and measure performance
- Exceptionally good levels of hazard control performance are achieved when hazard control is perceived as an important and integral part of planning, organization, direction and control decisions
- Hazard control management systems must be integrated into the main stream of all management functions
- There is usually an important difference between issued policy and procedure and what actually occurs
- Seldom is an activity as effectively managed as those responsible for it say it is

An audit is a part of the control function of management.

Evaluations of hazard control programs serve as appraisals of management performance in relation to issued policy and procedure. They are qualitative analyses of existing management systems, made to determine whether performance is effective and acceptable.

Above all, the goal of an evaluation is to constructively assist in improving the effectiveness of hazard control management systems. Although a survey of operations is significant in the evaluation process, unsafe physical conditions and practices are important principally as they may identify management systems which can be made more effective.

After having completed many evaluations of hazard control programs in a large variety of industries, it is apparent that there are elements common to all successful hazard control programs and that exceptionally favorable accident experience could not be achieved unless those elements were well managed.

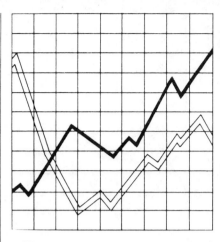

There are several studies which support this premise. Among them are an "Industrial Safety Study" by Thomas W. Planek, "Factors Apparently Affecting Injury Frequency in Eleven Matched Pairs of Companies" by Rollin H. Simonds and Yaghoub Shafai-Sahrai and "Factors in Successful Occupational Safety Programs" by Alexander Cohen. Similar conclusions were drawn in all of these papers. That to which I will principally refer was written by Alexander Cohen.

Mr. Cohen commenced his paper with two questions: What are the critical determinants of a successful industrial safety program? Do safety programs of companies with consistently good safety performance reveal any unusual or distinctive features that may account for their success?

Mr. Cohen's study was a project of the National Institute of Occupational Safety & Health, begun in 1974. In the first phase of the study, questionnaire returns were analyzed from 42 pairs of companies in Wisconsin. Members of each pair were matched in industrial operation, work force size, and geographical sector within the state but differed by at least 2 to 1 in work injury experience as reported for 1972 and 1973. Cohen stated that:

Specifically, the NIOSH project found the following factors to be more evident in the low accident companies than in their high partners, and to be particularly prominent in the record-holding establishments:

1. Greater management concern and involvement in safety matters exist, as reflected by the rank and stature of the company's safety officer, regular inclusion of safety issues in plant meeting agenda, and personal inspections of work areas by a top plant official, in some instances on a near daily basis
2. There are more open, informal communications between workers and management and frequent everyday contacts between workers and supervisors on both safety and other job matters
3. There are tidier work areas with more orderly plant operations, better ventilation and lighting and lower noise levels
4. The work force has more older, married workers with longer job service and less absenteeism and turnover
5. There is more regard for the use and effectiveness of measures other than suspensions and dismissals in disciplining violators of safety rules, e.g., provisions for personal counseling in such matters
6. There is greater availability of recreational facilities for worker use during off-hour jobs
7. Greater efforts are made to involve worker families in campaigns for promoting safety consciousness both on and off the job
8. There are well defined selection, placement and job advancement procedures with opportunities for training in developing new skills

Our findings parallel those of Mr. Cohen.

When speaking of evaluations of the effectiveness of hazard control programs, two terms need definition—hazard control program and effectiveness.

A hazard control program is a set of management systems, designed to achieve hazard control goals. In measuring effectiveness, a determination would be made that activities do or do not achieve expected or intended results. Evaluations of hazard control effectiveness are qualitative, rather than quantitative. Judgments are required, in relation to expectations, in measuring the level of success.

It has been our experience in conducting audits of the effectiveness of hazard control programs that there are normally two such programs at a location—the one management thinks it has, and the one it really has. A degree of failure is implied if the hazard control program management really has is a great deal less than the one it thinks it has.

Whether audits are made by yourselves, by other hazard control practitioners in your companies or by consultants, evaluations would be made of a number of elements:

- **Management involvement**

Through an analysis of the many evaluations made by the staff at M&M Protection Consultants, one conclusion arose above all others.

Management obtains that accident experience which it establishes as acceptable—acceptable being the organization's perception of what management does.

If accident experience is considered to be unsatisfactory by management, my suggestion is that the difficult questions be asked. Has that experience resulted from the attitude management has displayed toward hazard control—displayed by what it does? Is the accident experience that which has been programmed—by inference?

It's impossible for there to be superior accident records if executive personnel do not display, by their actions, that they intend to have them.

Peter Drucker, a prominent writer and management consultant, has written: "What the boss does and says, his most casual remarks, his habits, even his mannerisms, tend to appear to his subordinates as calculated, planned and meaningful."

Hazard control programs fail when personnel perceive what management does as an indication that management really isn't interested in the control of hazards, not sufficiently interested to give active direction to hazard control activities through active involvement and participation.

Management is what management does. If what management does gives negative impressions, it's unlikely that a hazard control program will be successful.

In every entity which has achieved an outstanding safety record, the organization *knows* that upper level management is involved, is held accountable and holds subordinates accountable for their accident experience.

Involvement can be demonstrated and made visible in many ways—by regularly communicating on safety subjects, by serving as chairman of a

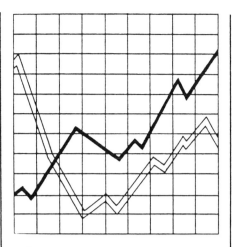

safety committee, by leading discussions of accident experience or other safety matters at staff meetings, but more importantly, by giving a strong emphasis to accountability.

- **Safety administration**

A professional safety staff is a necessity.

Hazard control programs are effective when safety staffing, time allocated to the function and resources are adequate. Appointing safety personnel with inadequate qualifications and then not providing for the necessary training will surely give a rather low status to the function of safety administration.

An organization will "read" the importance management intends to give to safety by appraising the qualifications of the persons assigned safety administration responsibilities, and their reporting place in the management structure.

It has not been possible for us to establish a pattern of safety organization typical to all entities with successful hazard control programs. More apparent is the level of capability and utilization of professional safety personnel, whose views are sought by upper level executives. They are considered to be a part of management and it's made very clear that they have access to top management. They have the capability to face real problems, to define solutions and to present their recommendations in a language that management can understand. They are asked to become a part of the decision making as respects hazard control in new facilities, in principal alterations and in equipment design and purchasing.

If the safety director's position is treated as insignificant, management instructs the organization that safety is insignificant.

- **Safety committees**

Depending on size of operations, union contract requirements and safety organization history, a variety of safety committee structures exists. If they are not effective, their existence can become a detriment to good safety management. These procedures should be followed if it's intended that a safety committee not be effective: schedule meetings for a particular time and place and cancel at least half of them; be sure that at least half of the management representatives who are to attend find other things to do that they consider more important; very early in the meeting, have the remaining management members called away on urgent matters; insist that the committee get bogged down in discussions of insignificant detail so that there is no time to make proposals of importance; if any recommendations are made by the committee, ignore them for at least two years; create the atmosphere that all matters pertaining to safety, no matter how trivial, will be handled by the safety committee, thus guaranteeing its ineffectiveness and the avoidance of responsibility for safety at all levels of management.

- **Supervisory participation**

Supervisors will do what they perceive to be important to their bosses. And if their bosses, by what they do, convey to supervisors a relative unimportance of hazard control, be assured that most supervisors will so respond. If supervisors are not made aware of their responsibility and held accountable for the control of hazards in their areas of operations, certainly, failure will result.

If you don't give supervisors safety training, if you don't impress upon them the importance of hazard control in their day-to-day supervision, if they are not made to understand that they are an important key to the prevention of accidents, you can't very well expect them to take the subject seriously. Nor will they.

It isn't enough, though, just to conduct a supervisor's safety training program. Management has to follow through by seeing that the content of the program is applied.

In organizations with successful safety programs, supervisory training in safety fundamentals reflects what is actually expected of supervisors, as displayed by what upper level management does.

- **Selection & training of employees**

Employees need to know, very soon after employment, that they have entered an organization that gives importance to safe performance.

It's typical for there to be a very thorough indoctrination program which gives emphasis to safety. As new employees pass through the indoctrination procedure, and eventually are assigned to a supervisor, they are very quickly able to evaluate the level of safety expected and whether suggestions for improved safety would be given serious consideration. Allow me to emphasize again the importance of the supervisor. Whatever example he sets will be followed by the new employee.

Consider this situation. Rather early in an evaluation of safety activities, an industrial relations director reviewed in great detail with the auditor the written indoctrination program for new employees. During the course of the evaluation, an interview was arranged, at random, with an employee who had been in the shop for about three months. In attempting to determine what he thought of the indoctrination program, he responded—what indoctrination program? This employee had bid up to his third job, was unaware of the company's indoctrination program, complained that he never saw his supervisor and didn't know how to get anyone to pay attention to gearbox covers that had been removed and not been replaced.

Under those circumstances, it would be impossible for management to convince new employees that the organization was serious about running a safe shop.

Surely, the available labor market, EEOC and union requirements present some difficulties in the selection of applicants that are mentally and physically capable of performing the tasks to which they are to be assigned. Those difficulties should not result in the abandonment of employment prerogatives which would surely result in a higher than wanted accident experience.

Unfortunately, employee safety training is much talked and written about, but often it's poorly done. You can't very well expect employees to follow safe practices if they have not been instructed in the procedures that are considered to be safe.

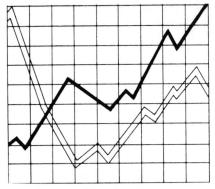

- **Preventive maintenance**

It's occurring a little too frequently, during our evaluations of safety program management, that the written programs of maintenance departments are found to be not nearly as effective, in relation to the control of hazards, as maintenance superintendents would lead us to believe. As discussions are held with the maintenance superintendent, we usually listen to a description of what is considered to be an excellent maintenance program, and on paper, it is.

Assume that you are making an audit of hazard control effectiveness, and that during your plant tour, you observe many unsafe conditions. You seek our a supervisor to ask why work orders aren't being sent to the maintenance department to have unsafe conditions corrected. And the response is: "We don't do that anymore. Safety work orders are the last priority for the maintenance department." You subsequently determine, through your audit procedures, that there are so many work orders over six months old pertaining to safety that discussions with maintenance department personnel become embarrassing. But, the maintenance program, on paper, was supposed to prevent that sort of thing from happening.

Management must maintain a safe environment if employees are to believe that a safety program is to be taken seriously.

- **Human factors— industrial engineering**

This subject, is generally given inadequate attention. It is our view that there is much talk about human factors engineering, but not quite enough action. Every aspect of the environment would require consideration and audits would logically include such items as machine design, illumination, heat and cold, humidity, dust and noise.

- **Control of health hazards**

It has been much easier to write occupational health control procedures than to obtain compliance with them. Audits would explore management systems to control occupational health hazards actually in place, in relation to what had been written, or not written. Ineffectiveness often results when the form of an occupational health program is given more importance than its substance.

This is a very broad subject and rather complex. Each industry has its own requirements for identification, evaluation and control of occupational health hazards, which have to be given a much greater attention because of OSHA.

- **Accident reporting & investigation**

Lack of training and follow-through will surely promote inadequate accident investigation. You can't very well expect supervisors to perform an effective job of accident investigation if they don't know how. Nor can management expect supervisors to treat the subject seriously if shallow investigations are accepted and treated as the norm, thus allowing real accident causes to go uncorrected.

Accident investigation is an exceptionally important part of a good hazard control program. Many significant messages are conveyed through good, or not so good, accident investigation procedures. Also, it's difficult to achieve effectiveness in other aspects of safety programming if corrective action is not taken to eliminate the causes of accidents which have occurred.

It takes a great deal of training and follow-through by safety professionals and management to achieve a high level of sophistication in accident reporting and investigation.

- **Recording & use of accident data**

Mention was previously made of the importance of upper level executives holding poor performers accountable for their records and for obtaining improvement. Sound accident data must be available if the executives involved are to make sound judgments. Much is made of accident data in the audits completed by our staff. The quality of these functions provides good indicators of the real expectations of management.

- **Measurement of progress**

For this subject, reviews would be made of goal setting, statistics available that would indicate that goals are or are not being attained, and the use of cost trend data.

If there is to be an effective accountability system, there has to be a base of measurement. Obviously, there are a variety of measurements and their appropriateness and utilization would be evaluated.

- **Inspections**

Effective self-inspections have many purposes, one of the most important being that they display management's determination that unsafe conditions and practices will be observed and corrected. There are no inspections more effective than those made by senior executives. It's amazing how much scurrying about and beneficial results those inspections produce.

In any case, internally conducted inspection programs can easily be measured for effectiveness. If a plant tour indicates many obvious, unsafe practices and conditions, it's doubtful that inspection programs are effective. Also, if the results of inspections aren't given appropriate attention, they become valueless.

- **Medical and first aid**

In-plant medical and first aid facilities are superior—not good, superior—in those organizations with outstanding safety records.

Of course, medical and first aid programs have improved greatly with the coming of OSHA. The proper treatment of industrial injuries and illnesses, immediately, is the base of the audit process.

- **Safety rules**

Establishing safe procedures, making certain that they are understood, and enforcing them become more important in our evaluations of safety activities.

Those organizations which have written good, sound, safe practice rules can have their effect diminished, and destroyed, if supervision does not convey, by what is done, that the rules are to be applied.

More often than not, the following, quoted from one of our evaluation reports, is typical:

  ☐ There are safety rules and new employees are given copies of them. Unfortunately, they are not reviewed extensively in the indoctrination procedure. Nor do supervisors use them as a part of their training in all departments. Quite a few employees had forgotten that the safety rules existed.

  ☐ Although safety rules seem adequate, on paper, a review of accident investigation reports indicates that many accidents occur because of direct violation of the written rules.

  ☐ Disciplinary action concerning safe procedures has been made important by some department managers but not by others. As an example, employees in some departments are allowed to work without eye protection while it is vigorously enforced in others.

- **Maintaining interest**

In his article titled "Factors In Successful Occupational Safety Programs," Alexander Cohen stated that "Motivational and educational techniques, as a main program element, were rated the least important." Although maintaining interest in safety activities through literature, posters and other means is typically a part of hazard control programs, and effectiveness is reviewed in our audits, it is not given disproportionate prominence, and rightly so.

- **Emergency plans**

This subject pertains to the identification of potential and planning for such events as fire, explosion, electrical outage, storms, civil disturbances and medical emergencies.

Depending on the nature of operations, a review is made of the existence and adequacy of emergency plans and whether they are kept updated as to personnel assignments and through practice drills.

- **OSHA compliance program**

Our review includes a determination of the steps taken to comply with OSHA, whether priorities have been established regarding known OSHA violations, whether a thoroughly documented safety program exists that will be useful to demonstrate a positive safety activity to an OSHA inspector, compliance with OSHA requirements and how citations are handled.

Audits made to evaluate the effectiveness of hazard control programs should provide qualitative measures of success. They should define strengths and weaknesses of hazard control program activity and relate to management systems in place, those that are actually functioning, and their quality.

Considering other measures as well, you would know that your hazard control program is effective if, subjectively, the results of an audit were satisfactory.

It was proposed earlier, and is again emphasized that the best approach is to use several measurement systems to evaluate the quality of hazard control management. ✚

Reprinted from the Fall 1979 issue of Viewpoint, *the Marsh & McLennan quarterly.*

References

1. Suggestive, Predictive, Decisive and Systemic Measurements: C. West Churchman: Paper delivered at an Industrial Safety Performance Symposium, National Safety Council, 1970.
2. The Process of Hazard Control: Robert J. Firenze: Kendall/Hunt Publishing Company, Dubuque, Iowa.
3. Loss Control Management: Bird and Loftus: Institute Press, Loganville, Georgia.
4. Applying Measurement Concepts to the Appraisal of Safety Performance: William E. Tarrants: Journal of the American Society of Safety Engineers, May 1965.
5. Towards More Effective Safety Measurement Systems: Herbert H. Jacobs: Journal of Safety Research, September 1970, National Safety Council.
6. Method of Recording and Measuring Work Injury Experience Z16.1: American National Standards Institute, N.Y.
7. What Every Employer Needs to Know About OSHA Recordkeeping: U.S. Department of Labor, Bureau of Labor Statistics, 1978, Report 412.3.
8. Techniques of Safety Management: Daniel C. Petersen: McGraw-Hill Book Company, N.Y.
9. Safety Management: Grimaldi and Simonds: Richard D. Irwin, Inc., Homewood, Illinois.
10. Industrial Safety Study: Thomas W. Planek: National Safety News, August 1967.
11. Factors Apparently Affecting Injury Frequency In Eleven Matched Pairs of Companies: Rollin H. Simonds and Yaghoub S. Shafai-Sahrai: Journal of Safety Research, September 1977.
12. Factors In Successful Occupational Safety Programs: Alexander Cohen: Journal of Safety Research, December 1977.

"... we felt that it would be desirable to have the top member of management responsible for the design and completion of major construction projects to become directly involved."

Engineering project planner, a way to engineer out unsafe conditions

by Donald J. Eckenfelder, CSP,
Corporate Safety Manager,
Chesebrough-Pond's Inc.,
Greenwich, Connecticut,
and
Charles E. Zaledonis, CSP,
Safety Manager—Rahway,
Merck & Co., Inc.,
Rahway, New Jersey

ABSTRACT. The authors have developed and are refining a technique for coping with the problem of overlooking design considerations from the standpoint of safety in implementing a new construction project. Mistakes in design run the gamut from minor "head-knockers" which can be corrected for a few dollars to major design mistakes which may cost thousands of dollars.

It is obvious and irrefutable that the best way to avoid unsafe conditions or physical hazards is to see to it that they are not built in when a facility is created. Unfortunately, in many cases, even conscientious people who recognize this fact fail to take proper precautions to prevent these unwanted conditions.

The Safety Professional's position

For years safety professionals have taken the position that somewhere between a majority and 80% of the accidents which produce industrial injuries are the results of unsafe acts of people. Even if this is true, still left is a rather substantial number of incidents that can be dealt with by providing safer facilities. Substantiating the need for safety in design are recent analyses using the OSHA recordkeeping system and involving a sample of several thousand people, which have indicated that unsafe conditions are probably involved in more than 50% of all accidents.

The study took place over a three-year period (1972-1974). The facilities involved have a comprehensive preventative-maintenance program and most people would describe the maintenance as well above any industry average (if it were possible to establish an average for such a subjective subject).

About 3,500 employees were involved. The incidence rate was below 10 for the three-year period. Supervisors filling out accident reports were asked to indicate if the cause was an unsafe condition, unsafe act, or if both contributed to the mishap. Their work was spot-checked by a safety professional. It was found to be reasonably accurate.

An 80-20 breakdown is obviously myth to anyone who has investigated occupational injuries and illnesses. In at least some of the cases, both unsafe conditions and acts are present together. Our study showed above 25% of the time; hence, no such simplified analysis is possible.

The data studied showed an unsafe condition being necessary to produce the untoward event in 51.44% of the cases studied where clear definition was possible. In all cases where the cause was unclear, the information was not tabulated or counted at all.

These conclusions provide all the more reason to address ourselves to the safe design of facilities.

In considering the problems of designing safe facilities, it is readily accepted that most large organizations involved in major construction projects over the years have made mistakes in planning that would have or did permit the creation of unsafe conditions. These mistakes have run the gamut from minor "head knockers," capable of being corrected for a few dollars, to major design errors on such things as automatic fire-protection systems. Costs in the latter kinds of cases may run in the tens of thousands of dollars to bring about the level of protection sought.

(continued on next page)

about the authors

Donald J. Eckenfelder, Corporate Safety Manager for Chesebrough-Pond's, Inc., is Past President of the Society's New Jersey Chapter. Prior to his present position he was employed as Safety Manager with Merck & Co., Inc., and earlier as a Fire Protection Engineer with Factory Mutual. He earned a bachelor of science degree in chemical engineering from Lafayette College in 1963.

A Certified Safety Professional, he has designed and taught a safety engineering course at Newark College of Engineering Graduate School. He is a qualified instructor for the ASSE-T Course and is a member of the Middlesex Community College Ad Hoc Committee to Implement OSHA Instruction Grant and Instructor In Program. He is ASSE Task Force Chairman, Inter-Society Relations, under the V.P. for Professional Affairs. He is a member of A.I.Ch.E., S.F.P.E., and A.I.H.A.

Charles E. Zaledonis, Safety Manager, Merck & Co., Inc., Rahway, N.J., is a member of the Society's New Jersey Chapter and currently is serving as Chairman of the Chapter's Public Relations Committee. Prior to his current position with his company, he served with the firm as Corporate Safety Engineer, Safety Supervisor, and Safety Engineer. Before that he was an Industrial Hygienist with the Pennsylvania Dept. of Health. He has a bachelor of science degree earned in 1964 from King's College.

A Certified Safety Professional, he is currently serving on the National Safety Council's Chemical Section Executive Committee. He is Chairman-elect of the Keystone Chapter, ACGIH, and is a member of the New Jersey Chapter of the American Industrial Hygiene Association.

```
                        PROJECT PLANNER

I.  PROJECT "LAYOUT"

    Exposure -

    A.  Will existing facilities present a danger to the new planned units
        by way of fire, explosion, noise, vapor, gas or mist hazards or will
        the new facilities endanger existing ones by any of the same
        potential hazards?

        1.  Are non-compatible materials segregated?

        2.  Are explosion venting cubicles directed to a safe location and/or
            are blast mats provided?

        3.  Are tanks or storage sheds, that will be used for flammable
            liquids, located near the property line or important plant
            buildings?  If yes:

            a.  Has separation by distance or rated fire walls been
                considered?

            b.  Has burying tanks been considered?  If so, a sump or
                other leak-detection system should be provided.

        4.  Are vents and other likely emission points of flammable vapors
            so located that build-up and ignition of a vapor cloud are
            minimal?  (See Standard Procedure).

        5.  Are fire walls needed to separate major areas of high value?

        6.  Has possible expansion been factored into separation consideration?

    Access -

    B.  Are roadways and walkways planned to provide safe access under normal
        maintenance and emergency conditions?

        1.  Are roadways looped to allow approach to all important facilities
            from at least two directions?

        2.  Have roadways, intersections, and parking facilities been planned
            to expedite the safe movement of motor vehicles in the Plant?
            An example might be the location of shipping and receiving facilities
            where outside commercial vehicles will not have to pass through
            production areas to reach their destination.

        3.  Are speed limit, stop, and other necessary traffic control signs to
            be provided?

        4.  Is line painting needed for pedestrian safety?
```

Figure 1.

For many years safety professionals in the nation have taken the position that somewhere between a majority and 80% of the accidents which produce industrial injuries are the results of unsafe acts of people...."

```
III. PROCESS
    Safety Testing -
    A.  Has safety testing been conducted by MSDPL to evaluate all process
        steps?
        1.  Did safety testing include raw materials?
        2.  Have suppliers provided Material Safety Data Sheets on their
            products?
        3.  Has MSDRL published reports describing their findings?
        4.  Where potentially hazardous reactions are known (Grignards,
            Friedel-Crafts, oxidations), has MSDPL supplied data on what
            types of control measures are necessary?
    Corrosion Testing -
    B.  Have corrosion tests been conducted by MSDPL and have corrosion test
        reports been issued to point out where special materials are needed
        for handling equipment?
    Health Testing -
    C.  Has health testing been conducted by MSDPL for all intermediates,
        by-products and products of the process being considered and are you
        aware of the results -- especially where special handling procedures,
        equipment or controls are required?
    Dust Handling -
    D.  Have dust explosion tests been conducted on all isolated, dry inter-
        mediates and products?
        1.  Have reports of these tests been published?
        2.  From the dust explosion test report mentioned in #1, are sufficient
            data known to enable you to construct the dry products handling
            equipment with adequate explosion venting protection?
        3.  Will the collectors, hoppers, bins, blenders, etc. be constructed
            strong enough to contain an explosion with the planned explosion
            venting capacity that is to be provided? Where explosion venting
            is not possible due to equipment configuration and/or location --
            explosion suppression systems may be required.
```

Figure 3.

On the other side of the coin, devices or hardware are sometimes called for in design which do little or nothing to reduce the potential frequency or severity of accidents. These mistakes are also expensive.

Fortunately, many of the types of errors mentioned above are subsequently recognized during construction and are corrected. However, the correction only comes about through the expenditure of additional unplanned time and dollars. And, some unsafe conditons doubtlessly remain undetected. Or they pose, at best, a moderate hazard and are impractical to correct.

In considering the problems associated with these design errors, we recognized that they are often committed despite the presence of a conscientious management which has a keen interest in people and property conservation. From our experiences, we found that even the efforts of highly competent engineers and safety professionals who recognize the need for careful engineering review of new projects to minimize hazards do not always prove a completely successful deterrent to these errors. Nor do seemingly thorough reviews of a project with engineering groups and insurance carriers provide, in our opinion, an adequate guarantee that the completed project will be as free as possible from hazards.

After cogitation on the problem, we come to the conclusion that better and more formal planning guides were needed to eliminate these design errors. We further concluded that even though there may be competent safety professionals and others on the staff who are aware of many of the considerations which must be observed in the design of safe facilities, much of responsibility for the project's safe design must rest with the project engineer. With the objective of helping project engineers to plan facilities in as safe a manner as possible, we set out to develop an "Engineering Project Planner."

The approach

In setting about our task, we decided that use of a comprehensive checklist would probably be the best tool to help cope with the problem of overlooking safety design considerations.

This approach was chosen even in recognition of the fact that checklists, no matter how comprehensive they may be, are seldom a panacea. We were thoroughly aware of the fact that even when all parties involved agree that conditions specified on the checklist are being met there is still opportunity for things to "fall through the cracks" during the various stages of design, fabrication, and installation.

Outweighing this disadvantage was the fact that we as safety professionals felt comfortable in using and sug-

```
II. BUILDING AND STRUCTURES
    Electrical -
    A.  Has the electrical classification of the area been determined?
        1.  Will positive pressure rooms be needed to house non-rated electrical
            equipment? Pressure monitoring devices, with alarms, may be
            required. (See Safety for more details).
        2.  Are switches, motors, instruments, lights, etc. to be used in
            hazardous areas of a type that is in accordance with the MCD
            Engineering Standard ME-2? If rated equipment is not available,
            pressure purging using ED131 may be required.
    Lighting -
    B.  Is area lighting adequate?
        1.  Are all exit signs lighted and light intensity acceptable?
        2.  Will emergency lighting systems be required? (See Standard Practices).
    Platforms -
    C.  Will all loading docks or any other elevated work platforms over 4 feet
        high have guard rails provided?
        1.  Where material must be lifted up to open-end platforms at high
            levels, is a movable handrail system provided?
        2.  Are elevators or other permanent lifting systems being considered
            over the use of fork trucks if large quantities of materials are
            to be raised to high levels?
    Ventilation -
    D.  Is area and/or specific ventilation necessary? (See Ventilation Standard).
```

Figure 2.

```
IV. Fire Protection
  B. Are water supplies adequate to meet additional demand?
    1. Has the new "maximum" water demand been calculated?
    2. Will new pumps and/or storage facilities be required?
    3. Can the existing fire water supply system provide sufficient
       and adequate pressure at the new use area, including hose streams?
    4. Are duplicate systems needed due to high values?
    5. Is the underground system adequate? Looping may be necessary.
    6. Are there sufficient hydrants of the proper type? Hydrant
       location must consider the hazard to be protected and generally
       should not be spaced more than 300 feet apart.
    7. Are sufficient sectional valves provided in the underground system?
Drainage -
  C. Is adequate drainage provided? Drainage calculations should include
     process, sprinkler and hose stream volumes.
    1. Is subdivision of areas adequate or needed to confine spread?
       Floor pitch should be at least 1%.
    2. Will drain capacity be a limiting factor on future expansions and
       changes?
    3. Have excess flow drains been provided where deluge systems are
       planned? For excess flow and any other drainage systems that
       might at some time carry flammable liquids, burning or not, care
       must be exercised to prevent creation of problems in other areas
       where this system is directed or connected.
Special Automatic Systems -
  D. Is there a need for any special protection that cannot be adequately
     handled by sprinklers?
    1. Should Halon protection be considered because of special occupancy?
    2. Should nitrogen inerting be considered for certain product handling
       equipment where fear of water damage from sprinklers is justified?
    3. Is fire or explosion suppression in equipment needed?
    4. Should fixed foam, CO₂ or dry chemical systems be considered?
```

Figure 4.

gesting the checklist, since it has proven to be such a valuable tool in our experiences in accident-prevention. Also, it seemed more desirable to provide this tool to the project engineer so that he could check his own work instead of being watched by a safety type.

```
V. UTILITY SERVICES
  Steam Service -
  A. Has steam generating equipment been installed in accordance with
     insurance carrier and governmental agency regulations and have the
     necessary approvals and/or certificates been received?
    1. Are steam lines insulated? Asbestos-free type only should be used.
    2. Are critical lines located where they are exposed to damage from
       a process or other type upset?
    3. Are there sufficient shutoffs and are they located where access is
       not difficult?
    4. Is the system to be looped or other methods to be used to insure
       steam to operations that might be damaged in the case of a sudden
       outage?
    5. Will service be monitored to detect problems with appropriate
       "interlocks" or alarms?
    6. Will noise-producing equipment be isolated or changed to reduce
       employee exposure?
  Air Service -
  B. Is the type of air supply system compatible with the service required?
     Breathing, instrument, process.
    1. Is moisture removal needed?
    2. Will air systems intended for multiple service be designed to
       prevent backup of contaminants into the breathing air and instrument
       air part of the system?
    3. Will externally lubricated compressors be used? Older type oil
       lubricated compressors should not be used for breathing air service.
       Only equipment capable of supplying CGA Grade "D" quality air should
       be used for breathing air systems.
    4. Will reserve tanks be provided to allow orderly shutdown of equipment
       in case of air supply failure?
    5. Will audible or other alarms be provided?
    6. Should critical process operations have a backup air supply source?
    7. Has the failure position of air service valves been reviewed?
```

Figure 5.

Before beginning formulation of the "Planner" we agreed that it was of paramount importance to reduce the safety planning duties of the project engineer to the simplest possible terms that would still provide an acceptable level of attention to these matters. We decided that this simplistic approach is necessary for proper usage of the "Planner." In this day and age most projects are terribly complex and require that the project engineer have a wide variety of skills and use them in complicated patterns, often on short notice. It is only fair that if one more burden is placed upon him it should be well thought out and be easily managed.

As we began to collect resource material to prepare this planner, we found that many other individuals and organizations had feelings obviously very similar to ours. The Manufacturing Chemists Association has published a booklet which assists in the evaluation of new projects from a safety standpoint. This booklet contains procedures developed by Monsanto which MCA reproduced. The A.I.Ch.E. makes available a publication which attemps to quantify the hazards associated with processes and in turn to identify adequate equipment, procedures, or facilities for use in minimizing them. Representatives of Union Carbide have given papers on the program they use for final plant evaluation following construction. It is comprehensive and involves safety personnel, line management, and persons who are involved in construction and design.

We are sure that there are many others who have considered and addressed this problem of how to assure that new facilities are as free from hazards as possible. The fact that this matter has received so much attention seems to add even more credibility to the effort.

The "Planner"

Continuing to emphasize the simple approach, we decided that an outline form using numbers and letters with key words to identify major areas would provide for the most effective use of the "Planner." We also reached the decision that, in order to achieve brevity, it would be desirable to reference company engineering standards and safety standards or requirements of outside agencies, such as those of OSHA or insurance companies. In this

"The final evaluation should be completed just before the facilities are turned over to operations. The evaluation team should include representatives of safety, operations, and engineering."

This checklist is provided for convenience to summarize on one sheet of paper the results of the survey using the Project Planner. If you have X's in boxes, that category will generally require an explanation and/or action. If **the X appears on a line** [either DNA (does not apply) or on a Yes or No answer], no action or follow-up is normally **required.**

		DNA	Yes	No
I.	A.	__	☐	__
	1.	__	__	☐
	2.	__	__	☐
	3.	__	☐	__
	a.	__	__	☐
	b.	__	__	☐
	4.	__	__	☐
	5.	__	☐	__
	6.	__	__	☐
	B.	__	__	☐
	1.	__	__	☐
	2.	__	__	☐
	3.	__	__	☐
	4.	__	☐	__
	C.	__	__	☐
	1.	__	__	☐
	2.	__	__	☐
	3.	__	☐	__
	D.	__	☐	__
	1.	__	☐	__
	2.	__	☐	__
	3.	__	__	☐
	E.	__	__	☐
	F.	__	☐	__
	1.	__	☐	__
	2.	__	☐	__
	3.	__	__	☐
	G.	__	__	☐
	1.	__	__	☐
	2.	__	__	☐
II.	A.	__	__	☐
	1.	__	☐	__
	2.	__	__	☐
	B.	__	__	☐
	1.	__	__	☐
	2.	__	☐	__
	C.	__	__	☐
	1.	__	__	☐
	2.	__	__	☐
	D.	__	☐	__
	1.	__	__	☐
	2.	__	__	☐

		DNA	Yes	No
II.	3.	__	☐	__
	E.	__	__	☐
	1.	__	__	☐
	2.	__	__	☐
	3.	__	__	☐
	4.	__	__	☐
	F.	__	__	☐
	1.	__	__	☐
	2.	__	__	☐
	3.	__	__	☐
	4.	__	__	☐
	5.	__	__	☐
	6.	__	__	☐
	7.	__	__	☐
	8.	__	__	☐
	G.	__	__	☐
	1.	__	__	☐
	2.	__	☐	__
	3.	__	__	☐
	4.	__	☐	__
	5.	__	__	☐
	H.	__	☐	__
	I.	__	__	☐
	J.	__	__	☐
	1.	__	☐	__
	2.	__	☐	__
	K.	__	☐	__
	L.	__	__	☐
	M.	__	__	☐
	N.	__	☐	__
	O.	__	☐	__
	P.	__	__	☐
	Q.	__	__	☐
	R.	__	☐	__
	S.	__	☐	__
	T.	__	__	☐
	1.	__	__	☐
	2.	__	__	☐
	U.	__	__	☐
	V.	__	☐	__
	W.	__	__	☐
	X.	__	__	☐

		DNA	Yes	No
III.	A.	__	__	☐
	1.	__	__	☐
	2.	__	__	☐
	3.	__	__	☐
	4.	__	__	☐
	B.	__	__	☐
	C.	__	__	☐
	D.	__	__	☐
	1.	__	__	☐
	2.	__	__	☐
	3.	__	__	☐
	E.	__	☐	__
	1.	__	__	☐
	2.	__	☐	__
	3.	__	__	☐
	F.	__	☐	__
	1.	__	__	☐
	2.	__	__	☐
	3.	__	☐	__
	4.	__	☐	__
	G.	__	☐	__
	1.	__	☐	__
	2.	__	__	☐
	3.	__	☐	__
	H.	__	☐	__
	1.	__	__	☐
	2.	__	__	☐
	3.	__	__	☐
	I.	__	__	☐
	1.	__	☐	__
	2.	__	☐	__
	3.	__	☐	__
	4.	__	☐	__
	5.	__	☐	__
	6.	__	☐	__
	J.	__	__	☐
	1.	__	__	☐
	2.	__	__	☐
	K.	__	__	☐
	L.	__	__	☐
	M.	__	__	☐
	N.	__	☐	__

		DNA	Yes	No
III.	1.	__	☐	__
	2.	__	__	☐
	O.	__	__	☐
	P.	__	__	☐
	1.	__	__	☐
	2.	__	☐	__
	3.	__	__	☐
	4.	__	☐	__
	Q.	__	☐	__
	1.	__	☐	__
	2.	__	☐	__
	R.	__	__	☐
	S.	__	☐	__
	1.	__	__	☐
	2.	__	__	☐
	T.	__	__	☐
	1.	__	__	☐
	2.	__	__	☐
	3.	__	__	☐
	4.	__	__	☐
	5.	__	__	☐
IV.	A.	__	__	☐
	1.	__	__	☐
	2.	__	__	☐
	3.	__	__	☐
	B.	__	__	☐
	1.	__	__	☐
	2.	__	☐	__
	3.	__	__	☐
	4.	__	☐	__
	5.	__	__	☐
	6.	__	__	☐
	7.	__	__	☐
	C.	__	__	☐
	1.	__	__	☐
	2.	__	☐	__
	3.	__	__	☐
	D.	__	☐	__
	1.	__	☐	__
	2.	__	☐	__
	3.	__	☐	__
	4.	__	☐	__

		DNA	Yes	No
IV.	E.	__	__	☐
	1.	__	__	☐
	2.	__	__	☐
	3.	__	__	☐
	4.	__	☐	__
	F.	__	☐	__
	1.	__	__	☐
	2.	__	__	☐
	3.	__	__	☐
	4.	__	__	☐
	G.	__	☐	__
	1.	__	__	☐
	a.	__	__	☐
	b.	__	__	☐
	c.	__	__	☐
	2.	__	☐	__
	a.	__	__	☐
	b.	__	☐	__
V.	A.	__	__	☐
	1.	__	__	☐
	2.	__	☐	__
	3.	__	__	☐
	4.	__	__	☐
	5.	__	__	☐
	6.	__	__	☐
	B.	__	__	☐
	1.	__	☐	__
	2.	__	__	☐
	3.	__	__	☐
	4.	__	__	☐
	5.	__	__	☐
	6.	__	☐	__
	7.	__	__	☐
	8.	__	☐	__
	9.	__	__	☐
	C.	__	__	☐
	D.	__	__	☐
	1.	__	__	☐
	2.	__	__	☐
	3.	__	__	☐
	4.	__	☐	__
	E.	__	☐	__

Figure 6.

way we kept the prime document as short as possible and still indicated where the user could find information needed to determine if that particular item noted in the "Planner" was in compliance or had been taken into consideration.

After drafting several outlines, we finally got to the

point where the "Planner" evolved into five basic categories: Project Layout, Buildings and Structures, Process, Fire Protection, and Utility Services. We had had a separate section for Industrial Hygiene but found that virtually all of the questions that applied to it were being asked under other sections.

There is overlap among the separate sections. We found it very difficult to draw clean lines between them. No effort was made to relate the sections to each other. The five sections exist primarily to create some type of breakdown. The strength of the "Planner" is in the individual items and not in the section breakdown. The sections will be refined and improved by ourselves or others.

The particular "Planner" that we developed addresses itself to construction projects for chemical operations. We do feel, however, that much of it has universal applicability and with slight modifications could be used to evaluate almost any type of project. The idea and format, we believe, are sound.

The "Planner" itself turned out to have a total of 208 items. The question-and-answer sheets numbered 16. For reference, we have included examples from each of the five sections and the "Project Planner Checklist." Figure 1 is from the Project Layout Section. Figure 2 is taken from the section on Building and Structures. A sample from the Process Section is provided by Figure 3. Figure 4 comes from the Fire Protection Section. An example from the Utility Services Section is provided by Figure 5.

A specific example of our procedures of referencing a standard procedure may be seen in Figure 1 under A.4. Similarly, an example may be seen in Figure 2 under A.2. of one of our references to specific engineering standards.

The Project Planner Checklist that we developed for use in summarizing the results of the "Planner" analysis is depicted in Figure 6. By giving this summary a quick glance, we believe that the user (or reader) can be provided with an idea of how things stand.

The format which was chosen in the development of the Project Planner Checklist has been used by Bob Rheinlander, Safety Manager for Calgon Corporation. It requires some action or follow-up if a mark is placed in any of the boxes. To use this technique, it is necessary to tailor the questions so that they can receive a simple positive or negative answer (you will note in looking at the reference sheets that we have used one kind of type face to ask the question and then used another kind to illuminate or comment on the area covered by the question).

Implementation

No matter how good a device our "Planner" turned out to be, we realized that its method of implementation would prove as the determinant of its actual usefulness. Toward that end, we felt that it would be desirable to have the top member of management responsible for the design and completion of major construction projects to become directly involved. We were convinced that the best way to accomplish this would be for him to communicate directly with the project engineer indicating the necessity to utilize the review list for analysis of the project from a safety and health standpoint.

This communique would also request a projected target date for completion of the "Planner" analysis. Copies of the requesting note would go to all concerned including the management for whom the project was being constructed and the local and corporate safety contact. It also would indicate that if any assistance was necessary it would be provided by other support groups, especially the local safety contact who so often is forgotten and left out of important decisions early in a project.

Unfortunately, what has been said up to now—although very positive—was just the beginning. Anyone who would sit back and look down the long path of a major construction project and expect everything to go smoothly has not followed too many projects to see what twists and turns occur along the way. Just as in so many other parts of our work, the key here is "follow-up."

To achieve a high level of confidence in seeing that things got done, we felt two scheduled reviews along the way were necessary to achieve our objective of building a productive, safe, and healthy facility. Hence, a formal mid-project and final evaluation were specified.

It is interesting to note that when a mid-project review has been used on large projects it has turned out to be an eye-opener to all concerned. It was found to be very surprising how differently people interpret the same statements and decisions made early during the design. The intent of this mid-project review is to identify these differing interpretations before irrevocable commitments are made which would not easily allow alterations to achieve the safe conditions desired.

The final evaluation should be completed just before the facilities are turned over to operations. The evaluation team should include representatives of safety, operations, and engineering.

Summary

It should be pointed out that the concept and tool covered in this paper are new not only to the reader but also to the people who created them. At this point in time the technique has not been used in its entirety on a major project. We have high hopes for it, but we recognize that in the course of breaking it in we will find deficiencies which require correction or revision. One of the things we have already thought about is the fact that some sections or individual questions are of high import and involve substantial amounts of capital, whereas others are of a very minor nature.

We have considered the possibility of using some technique, such as hazard mode and effect analysis (HMEA), on each individual subject and attempting through the use of frequency and severity evaluation to develop criticalities for the individual subjects and to reflect this in some fashion in the "Planner." We feel the concept is worthy of a trial and that it will meet with sufficient success to at least justify improvement, updating, and streamlining. Even though we have not yet used the "Planner" in a major project, we are anxious to share what we have with anyone who would be interested.

A complete copy of the "Planner" will be mailed to anyone who sends either author a self-addressed stamped envelope. The envelope should be of sufficient size to accommodate about 20 sheets of 8½ by 11 paper. We will collect requests for the "Planner" for approximately one month following publication of this issue of PS and then will mail all requested copies at once.

Comments or suggestions concerning the "Planner" also would be welcomed. –PS–

Warning label design

By Michael W. Riley, David J. Cochran, and John E. Deacy

The design of labels must be of concern to the safety professional. Two general categories of labels exist. The first category is instructive, performance or informational in nature, and the second category is warning or precautionary. In both categories workers who do not see, recognize, properly understand, or heed labels can potentially contribute to a significant safety problem. Thus safety personnel should be concerned with warning labeling.

Labeling conflicts

The good intentions of safety personnel can be subverted when the label design function is subordinated to the interests of sales, advertising, legal, materials, and finance managers. Complying with the mandatory labeling requirements of the United States Consumer Products Safety Commission and other regulatory agencies does not change the common law duty to warn. Compliance merely authorizes marketing. The number of agencies of government which specify labeling requirements is estimated to be approximately 125 and serious repercussions could occur if the regulations are not followed.

Merchantability, as specified by the United States Uniform Commercial Codes, implies safe packaging and proper labeling. The implied warranty of merchantability applies to both product and container; court decisions exist to support this contention.[1] A product's safety and safe packaging are synonymous since the physical packaging, its labeling, and its contents are characteristics which together determine the product's safety.[2] Instructions must include disposal warnings and cautions where hazardous products are involved.[3] An implied warranty can be made by over-emphasizing danger in one area and thereby implying safety in another.

Packaging functions not only provide security to goods during distribution, but also penetrate the consciousness of the buyer, to increase sales of the product, and to instruct the buyer to use it properly. If warnings are clear, reasonable, apparent, and come to the attention of the user, and if failure to comply results in an injury, theoretically the seller is provided with a defense.

Economics of labeling

Information added to a label voluntarily is normally intended to increase sales or to facilitate processing of the package. Labels added because of governmental requirements are intended for the consumer's benefit. In both cases, customers ultimately pay the cost.

A manufacturer is also concerned about the economic impact of the label. Can a product conservatively labeled as hazardous effectively compete against a similar product not labeled as hazardous?

Since labels can act as warranties, labeling is directly linked with the cost of product liability claims. Consumers frequently buy a product based upon its perceived reliability and safety. Misperceptions can be boosted by assertions about product reliability and safety through labeling. The maker can then use increased profits through sales to offset the insurance necessary to cover added products liability claims.

It is believed by some that balancing the benefits and risks in marketing results in a complete and open disclosure of all facts. Contradictory data indicate that many firms knowingly do not have the "zero defects" philosophy of quality because of the economics. This implies that their expected products liability litigation costs and lost sales due to defective products are less than the expected costs for approaching "zero defects." A similar argument can be made about design defects.

Development of a label

Because of the conflicting influences in label design and the economics of label design, the safety professional may be in the best position to adequately suggest solutions and should be called upon to participate in label design decisions. Since warning labels are the labels of most concern, the remainder of the discussion relates primarily to them.

The process of developing a label should start with data accumulated at every stage of the development of the product; therefore, label development and product development should be simultaneous events. The acquisition of performance data, failure rates, hazard data, the keeping of detailed records, and corporate dedication to the systematic development of the label enhances the firm's ability to balance the risk associated with a product between design alternatives, quality control, warning labels and warranties; and to defend the product should litigation arise from a mishap in the use of the product. This effort is also expected to contribute to better quality control, which in itself is an aid in the prevention and defense of litigation.

The courts have often determined for the plaintiff (injured) partly because it was reasonable to assume that the customer's perception of the information supplied by the maker implied that the product was safe for the user's intended purpose—implied warranty. Manufacturers may not overpromote their products and downplay the dangers, even if the warnings are adequate and comply with regulations, because a consumer may be lulled into believing, through advertising, that the product was safe, despite the warning.[4]

In collecting the data necessary to develop an effective label, the product safety engineer must be aware that the social attitudes toward the contract of sale, between a manufacturer/seller and the ultimate consumer, involve human factors. The population of users has been educated through advertising to the notions of expertly designed products expressly fit for general applications, zero defects in production, complete honesty in contracts of sale, and infinite service life for many products. These social attitudes are fostered when companies bow to their marketing personnel and omit limitations from their disclaimers or otherwise overstate the product. The project engineer must address these complex variables using risk analysis in attempting to control the design and labeling process.

> **The label must point out the "good" . . . (and) limitations and potential hazards as well.**

The logical practice of dealing with some of the variability of production and consumption through the last possible alternative, the label, is obnoxious to sales personnel and confusing factor to the buyer. If the product manager resorts to the label or the price mechanism to control variability, the product may not compete in the market.

Through the balancing of the costs of product design, quality control, the education of the ultimate customer and litigation, a product is given a chance to survive in a competitive market. The label therefore must not only point out the "good" qualities of the product, but its limitations and potential hazards as well.

Dorris and Purswell[5] indicate that the human factors discipline has been slow to research the design of safe products and effective warnings. They believe that this lack of activity stems from two factors, the first factor being that there is little large-scale federal funding of basic product safety research from a behavioral perspective, and the work done as a consultant is usually kept confidential as proprietary information. The second factor is that behavioral problems involve phenomena which do not lend themselves to analysis such as factorial design and analysis of variance. New methodologies for evaluating the potential for injury are desperately needed.

The design of the physical layout of the message in a label, the alphanumeric characters, and pictorial designs have been extensively studied by human factors specialists. Most larger corporations make well documented systems, including reference materials available at a nominal price.

The following general outline is suggested for assembling the data necessary for label development (emphasis on safety):[6]

1. Obtain top management's commitment to the subjects of quality, reliability and product safety. Design choices must be made with the management being fully aware of certain risks of injury present in the design and that the choice resulted from balancing risk of injury, product utility, design alternatives and economic factors.[7]
2. Use fault-tree analysis and Failure Mode Effect Analysis (FMEA) to systematically identify, analyze and document failure modes and effects on system performance and personnel safety.
3. Test and evaluate the product under all foreseeable uses and keep detailed records. Simulate actual conditions if possible. (This simulation of product failures and personal injury may be difficult because of the costs of material or equipment and the potential for actual injury during the simulation.) Use these data to redesign or design safeguards and shields to isolate the user from the dangerous characteristics.
4. Conduct an evaluation of the trade-offs of product risks vs. the utility of the product using the standard of reasonableness.[7]
5. Identify the product so that the users know what they are dealing with. Provide enough information for the users to make judgments on alternate uses.
6. Evaluate and apply produc-

tion controls (quality control) to reduce defects to a specific low number commensurate with risk analysis.
7. Decide on the trade-offs between product safety and production costs which are to be converted to warning labels. Place the warning labels in a conspicuous place. Instruct new operators in safe use of the product. Supervise installations where necessary to reduce risks.
8. Set up a feedback mechanism to evaluate failure, defects, and injury resulting from use. React to all defects and keep detailed records.
9. Maintain "State of the Art" in technology. Retrofit or revise the design whenever technology can be used to increase social benefits at a balanced cost.
10. Keep tight reins on sales "pitches"—they may imply a warranty.
11. Design the label so as to include the warnings necessary to reasonably identify hazards that safety devices cannot cure.
12. Test the warning label for effectiveness to insure that the message intended penetrates the consciousness of all possible users, some of whom may speak a different language. Document the tests.

Greenstone[8] adds that the products safety engineer should be able to comment on the adequacy of tests and inspections, materials used, construction methods, and safeguards and precautions as compared to customs and practices within the industry. Scotti[9] suggests that the engineer must be able to provide a statement to the sufficiency and adequacy of safety features and efforts to anticipate foreseeable misuse. The approach by most consumer organizations on labels is a direct attack on industry, which, in the view of the consumer advocate, has stubbornly refused to inform the public in order to conceal the basic worthlessness of its product.[10] Surveys show that consumers want to know more about the products they use—especially food products.

Conclusion

Developing a warning label is a delicate balance between regulations, sales promotion, warranty statements, labeling costs, and product image. Designing a legally correct, clearly understandable, uncluttered, artistically pleasing, warning label is a difficult task. Such an effort requires a team effort with a safety professional providing much of the needed input. Criteria and methodologies for evaluating warning label effectiveness are needed. The safety professional must play an important role in such future activities in order to find the needed answers.

References
1. Sunquist, James L., *Politics and Policy: The Eisenhower, Kennedy and Johnson Years,* The Brookings Institute, Washington, 1969.
2. Wiley, Harvey W., *The History of a Crime Against the Food Law,* Harvey W. Wiley, Washington, p. 52, 1929.
3. Nadel, Mark V., *The Politics of Consumerism,* Indianapolis: Bobbs Merril Co., p. 9, 1971.
4. Elser, John T., "Prevention Cost vs. Settlement Costs," Proceedings of Product Liability Panel/76.
5. Dorris, Alan L. and Purswell, Jerry L, "Human Factors in Design of Effective Product Warnings," Proceedings of Human Factors Society—22nd Annual Meeting, Human Factors Society, Inc., 1978.
6. Stafford, Robert H., "Planning a Defense Before the Accident," 1977 Product Liability Prevention Conference, 22-24 Aug., 1977, Hasbrook Heights, New Jersey. Newark: New Jersey Institute of Technology, p. 205.
7. Weinstein, A. S., Piehler, H. R., Twerski, A. D., and Donaher, W. A., *Products Liability and a Reasonably Safe Product,* New York: Wiley and Sons, p. 43, 1978.
8. Greenstone, Herbert E., "What an Attorney Expects of an Expert Witness," 1977 Product Liability Prevention Conference, 22-24 Aug., 1977, Hasbrook Heights, New Jersey, Newark: New Jersey Institute of Technology, p. 177.
9. Scotti, Marie, "Guidelines for Company Safety and Product Loss Prevention," 1977 Product Liability Prevention Conference, 22-24 Aug., 1977, Hasbrook Heights, New Jersey, Newark: New Jersey Institute of Technology, p. 149.
10. Seligsohn, "Food Label Fever Grips Washington," *Food Engineering,* Vol. 50, No. 7, July 1978, p. 20.

New dimensions in the tortious failure to warn

by Harry M. Philo

On January 15, 1969, the National Machine Tool Builders Association mailed out Accident Prevention and Safety Bulletin No. 69-4 which said in relevant part:

"The Metal Forming Subcommittee of the Association's Accident Prevention and Safety Committee has concluded its evaluation of a warning sign for use by this segment of the Machine Tool Industry. The Committee believes that the use of *warning signs can be of prime importance in helping to eliminate accidents. The proper use of warning signs can also reduce possible exposure to product liability suits.*

"The signs have been designed to afford adequate and readable instructions to the operator of the machine on which it is installed.

"Embossing of the sign provides a three-dimensional appearance to improve readability. The three colors used, black, white and red, are in accordance with guidelines found in U.S.A.S.I. Standard Z-35." (Emphasis added.)

The Bulletin is particularly illustrative of many points this article will seek to make:

1) The year 1969 represents the approximate turning point in the recognition of what products liability means to product safety; the social purpose of tort law is accident prevention. The tort law, which historically had given business a legal license to kill and maim, was beginning to eliminate that privilege in sufficient degree as to be recognized by a major association of manufacturers.

2) The preparation of a warning is very sophisticated business.

3) Warnings are of prime importance to the elimination of accidental injuries and deaths.

4) Warnings require standards and conformance to standards, and standards were in existence.

5) Warnings must be on the product or machine itself.

6) The low level of safety consciousness of industry generally required that an industry association remedy the failure of the individual companies and even produce subsequent modification signs. The signs, which were aluminum and which cost from 40 cents to 70 cents, were intended for placement on machines which sold in six and seven figures.

The trial lawyer's role

The design engineer *who has done a proper failure mode and effect analysis* and recognized the hazardous uses of a product, evaluated the risk of injury and understood the gravity of that risk, and further determined that the risk was unreasonable and unacceptable often is in the

position that neither the hazard nor the risk can be eliminated reasonably by guarding, interlocking, fail-safe-ing or nontoxic substitutes. There remains a risk of serious injury or death. Reasonable prudence requires a warning which describes the hazard, alerts of the grave risk, states the safety engineering alternative and often details emergency treatment, if injury occurs. Let us then examine the purpose, the history of, and the quality of necessary warnings.

We must do this because no one else has. Just as trial lawyers are forced every day to redesign products for safety and discover the etiology of chemical insult, so too are we forced to bring together the work of all the various disciplines concerned wtih warnings and their adequacy.

The prudent preparation of the warnings so necessary for societal safety combines the expertise of:

1) The safety engineer who is trained in the recognition of hazards and risks through use of the tools of the safety profession;

2) The psychologist or the human factors engineer concerned additionally with understanding the lack of hazard and risk recognition of those with simple "common sense";

3) The semanticist who understands that the intended meaning of a writer does not necessarily coincide with what is the understanding of the reader; and

4) The graphic artist trained to recognize the color combinations, signal words, kind of type, size of type, attaching mechanisms, etc., to meet the purpose of a warning.

While trial lawyers often have the luxury of hindsight, the prudent design engineer must either combine the aforementioned skills or employ consultants with the necessary sophistication.

When the trial lawyer seeks to understand the history, purpose, and scope of warnings, it becomes obvious that neither lawyers, judges, nor the law are as yet with it when it comes to the law and warnings.

History

Accident prevention signs probably have been used longer than any other safety device or piece of safety equipment. In fact, they were probably used even before 1450, when Scotland enacted the first statutes governing product hazards. The first American standards governing warning signs appeared in 1914 in a book titled *Universal Safety Standards*. One chapter or section was called "Standards For Danger And Safety Signs." The concepts, approaches, and specifications are similar to what we have today in the American National Standards Institute's Z35.1 Standards.

Many American industries have made some good faith efforts to prevent accidental injuries and deaths by use of warnings. The only industries to deliberately violate everything that has been learned historically about warnings are the pharmaceutical and cosmetics industries which hide and obscure the necessary information about the dangers or inadequacies of their drugs and cosmetics. They, as one might expect, have been given the broadest license to kill and maim by backward common law decisions which grant them protection against paying damages for much of their culpability.

We have long had standard specifications for Industrial Accident Prevention Signs. The Manufacturing Chemists Association first published its *Guide to Precautionary Labeling of Hazardous Chemicals* 36 years ago, and it became an ANSI Stan-

> **Reasonable prudence requires a warning which describes the hazard . . .**

dard in 1976. The *Uniform Traffic Control Device Manual* of the federal government, detailing the color, shape, placement, and purpose of every highway device, has been incorporated by reference in the statutes of each state for decades. It includes the warrants for and details of railroad grade crossing warnings. It has for more than 15 years described the kind and quality of highway construction warnings. The Federal Caustic Act of 1927 covered mainly the labeling of some caustic corrosive chemicals intended for household use. The Pure Food, Drug and Cosmetic Act of 1938 was passed with the pretense of requiring adequate warnings on the labels of all drugs. The Federal Insecticide, Fungicide and Rodenticide Law of 1947 mandated labels with warning statements on economic poisons. The Interstate Commerce Commission (ICC) by regulations requires particular warning labels on explosives and dangerous materials when they are shipped in interstate commerce.

The United Nations, through UNESCO and the International Labour Organization, has been attempting with some success since 1955 to get international signal words and symbols with universal recognition.

The Consumer Product Safety Commission (CPSC) has attempted to mandate and has also suggested warnings for such diverse products as water slides and football helmets, among other things. It has jurisdiction over conformance with the Federal Hazardous Substances Act which governs warnings on products which may foreseeably be used in the household.

When one understands the extremely low level of safety consciousness in the United States and the absurd level of attempted warnings which have been given historically, it is readily recognizable that there is presently a climate which underestimates the tool of warnings as a major accident prevention device. There is, however, an unconscious recognition that exhortations to "be careful," "operate your machine safely," and "drive safely" have caused more injuries and deaths than they ever prevented. Warnings constitute a much higher level of sophistication in communication. A failure in communication is near-conclusive evidence of negligent design.

We have had safety color codes for marking physical hazards since 1945. There has been a separate code for warning tags since 1966. There has been national recognition of the safety movement in the United States since the first national conference on industrial safety was held in 1912. All participants in the safety movement at that time used safety signs.

We have long had scores of companies producing safety signs and decals and buying booths at the scores of conferences and seminars.

The purposes of warnings
In examining the purposes of warn-

> *... there is presently a climate which underestimates the tool of warnings ...*

ings, it may first be desirable to detail what warnings are not.

1) In safety engineering a warning is never enough if one can do better. If the hazard can be eliminated reasonably, then it must be eliminated. Zero danger results. If the risk can be eliminated reasonably, then it must be eliminated. Zero danger results. If the hazard can be reduced reasonably, then it must be reduced. If the risk can be reduced reasonably, then it must be reduced.

The next step in safety engineering is instructions for safe and effective use and warnings against risks that may cause injury and death.

2) An inadequate warning is no warning at all.

3) A warning is not a statement for the sole purpose of avoiding liability. Disclaimers seek to avoid liability. Warnings are aimed at preventing accidental injury and death. Obviously, misrepresentations that interfere with warnings should not be used.

An important concept in safety engineering is that of the banned product. The remaining risk is so great that no warning is sufficient. The product must be banned by

statute or regulation. If not, the common law court can find it should have been banned.

Proper warnings can fulfill any or all of a number of purposes.

1) An *informed choice* for the *entity in charge of the environment*—the employer, the farmer, the parent, the athletic director, the equipment purchaser is alerted to the fact that a manufacturer or one initiating an activity could not get zero risk, and there remains a risk of serious injury or death so that the new entity having been adequately informed of the gravity of the risk may decline the risk entirely.

• The employer may not buy the machine.
• The farmer may not buy the switch that has not been explosion-proofed.
• The mother may seek a nontoxic substitute.
• The athletic director may abandon football rather than have the players play with inadequate helmets to protect against brain damage and spinal cord injury.

2) An *exhortation* with the intensity required to alert the *entity in charge of the environment* of the nature of the hazard—the gravity of the risk so that the hazard and risk may be further reduced.

• The employer may use die guards to limit access when the machine builder did not guard point of operation.
• The farmer may use a big round bale clamp on the front end loader when loading such bales.
• The mother may prevent access by the child to the poisonous household chemical.
• The employer may provide chemical goggles and emergency equipment to workers working with aqua ammonia.
• The coach may start teaching football play in a way to minimize

the number and severity of blows to the head.

3) *An informed choice for the user.* The *user* may well choose not to undertake the risk at all—zero danger/zero injury.

● The young woman may decline taking birth control pills with the very high probability of death, thrombotic stroke, blood clots, cancer, or deformed children.

● The farmer may decline to use the particular pesticide.

● The student may forego use of the trampoline.

● The car purchaser may want more fireworthiness than given by some models such as the Pinto or more crashworthiness than some motorcycles.

4) An *exhortation to the user* with the intensity required so that the user will take further steps to minimize the hazard and risk.

● The worker may not weld on the empty drum.

● The vehicle owner will not crank the engine while spraying quick-start into the carburetor.

● The motorcyclist may not drive the motorcycle up a hill with a passenger in the first 500 miles.

● The window washer may use a safety belt and independent safety line rather than rely upon the powered scaffolding.

● The painter may not use a ladder to paint from and insist upon a scaffold with guard rails.

5) A *reminder* to someone who already knows:

A very important purpose of a warning is to remind one who already knows the danger but whose attention has been diverted or arrested by competing circumstances, competing tasks or heat, light, noise, etc.

6) To *assure safe disposal* of containers whose contents were toxic, corrosive, caustic, etc.

7) To *minimize the injury*. A major purpose of a warning is the minimizing of the extent of injury, i.e.:

● There may not be movement of the paralyzed person without a spine board.

● The poison insult may be lessened with an antidote.

● There may be immediate massive flushing of the *eyes* where there has been alkali splashed into the eyes.

● The exposure to vapor may be minimized.

● The spillage may not cause fire.

8) *To assist a recall campaign.*

. . . a warning is to remind one who already knows the danger . . .

Warnings subsequent to sale are becoming commonplace for all the above reasons. They are regularly used by industry to hasten recalls. The Department of Transportation, the Food and Drug Administration, and the Consumer Product Safety Commission have all been part of warning campaigns in the last 15 years. The *Product Liability Portfolio* by Man and Manager Inc., a major industry management advisory service, has helpful material for the trial lawyer seeking to find evidence of reasonable prudence in these kinds of policies and procedures.

There are, of course, times when reasonable prudence might be met by warning campaigns less than recall. This is a major question in the conduct of Ford in its park-to-reverse transmission problems.

The backwardness of the law

The A.S.A. Z35.1 Code which was adopted 40 years ago pursuant to directions from the American Standards Association and the National Safety Council as to their scope:

"Design, application and use of warning signs . . . intended to indicate and insofar as possible to define specific hazards of a nature such that failure to so designate them may cause or tend to cause accidental injury to workers or the public or both.

If there is a failure to warn, then in every instance that failure is a cause of the injury or death complained of unless the injured person was attempting suicide, was blind or blindly intoxicated, or was intentionally undertaking an unreasonable risk. In every instance where there was a failure to warn reasonable prudence would have alerted the defendant that harm could result unless one of the exceptions existed. This is so because the activity engaged in or the use of a product was unreasonably risky. The *hazard* presenting the unreasonable or unacceptable risk *caused* the injury. The hazard and risk could only be made acceptable with a warning.

The law has put the cart before the horse. It has been deficient in several respects and ill serves society as heretofore expounded.

Causation. It is absurd to have a jury groping and speculating on failure-to-warn causation. If a product can only be made reasonably safe with a warning, then it is not the failure to warn which causes injury. It is the unreasonable exposure to the hazard which is a cause of the injury. Unreasonable exposure occurs with a combination of hazard (an objective condition or changing set of circumstances) and risk (probability of a particular untoward result). All the erroneous jury instructions need to be corrected to reflect the true situation and eliminate the artificial testimony, "I would have obeyed that warning."

Burden of proof of causation. It is equally absurd to place the burden upon a plaintiff who has been injured after having been subjected to an unreasonable exposure to a hazard.

Once the plaintiff proves that a warning was necessary, the only logical and just method of handling this is to allow the defendant to prove that:

1) The defendant gave an adequate warning;

2) The plaintiff was attempting suicide;

3) The plaintiff was consciously exposing himself or herself to a hazard as a thrill seeker knowing the gravity of the particular risk; or

4) The plantiff, either because of a blind drunkenness or mental in-

capacity to be helped by a warning, could have not been exhorted or reminded in a way to prevent injury, and there is no other way that the defendant, acting reasonably, could have prevented the particular unreasonable exposure.

Texas and Indiana are the only jurisdictions which have definitively shifted this burden.

Ordinary Consumer Expectations. The safety and design engineering principle—so crucial for accident prevention—*Any Risk of Serious Injury or Death Is Always Unreasonable and Always Unacceptable If Reasonable Accident Prevention Methods Would Eliminate or Minimize It*—does not limit protection to the ordinary consumer, but rather to anyone who could reasonably foreseeably be injured because of lack of warning. It is in the economic and moral interest of society to protect "the worst damn fool" when that can be done. The test then is simply scientific and economic feasibility of warnings.

Limitations on supplier's knowledge of danger. Just as strict liability is a shortcut based upon social reality (it is *fault* if you sell a defective product), so it is also necessary for the law to understand that a prudent failure mode and effect analysis in the design of every product will document every way the product can fail (or the inadequacies of testing), and the law should conclusively presume the necessary knowledge by a defendant of hazard and risk, and lack of risk recognition on the parts of users and the safety engineering alternative.

No duty to warn patient. Because of a succinctly stated special privilege, drug dealers have no duty to warn the patient if the physician has been adequately warned. It denies patients an informed choice and belies the reality that injury prevention can be greatly enhanced if patients know the ways to avoid unsafe drugs, avoid injury by understanding contraindications, and minimize injury with correct dosage, etc. It is constitutionally unreasonable privilege to give immunity to drug dealers dispersing such products as birth control pills which emphasize patient choice of contraception rather than doctor decision.

No duty if the danger is open and obvious. If there has been an expression in law comparable to, "The moon is made of green cheese," it is, "The danger is open and obvious." It was a stupid expression when first opined in *Campo v. Scofield,* and it was equally stupid in each subsequent expression. When Mrs. Campo put her hand into the jaws of the onion topping machine, one hundred thousand workers were putting their hands into the jaws of a machine some eight hundred times that day, and not one of them considered it an unreasonable risk. The question is not whether there is a recognizable hazard, or whether the user will recognize the percentage chance of injury, or whether the user will recognize the gravity of the risk or recognize the safety engineering alternative. The real question is: "Has the risk of injury been reduced as much as reasonably possible?"

If it has, the defendant should be relieved of liability; if it has not, the defendant should remain liable for that failure which results in injury.

Modification of the product. If there is anything that is foreseeable to a designer it is that the product will be modified. If there is anything foreseeable in safety engineering, it is that employers will remove guards. This has been important knowledge for at least 100 years. Products are modified proportionate to the inadequacy of the design. Guards are removed because of negligent design. Negligent design includes the failure to warn against modifications and its dangers.

Hiding subsequent warnings from the jury. There is a general maxim in science, philosophy, and law that one has adopted the wrong principle if the principle has multiple exceptions. Such is the immoral rule that a tortfeasor can hide its subsequent warnings from the jury in certain instances. The general rule is subject

> **It is in the economic and moral interest of society to protect "the worst damn fool . . ."**

to exceptions which make the subsequent warnings admissible for showing control, scientific feasibility, economic feasibility, impeachment causation, prior knowledge and modification by others. Maine recently has corrected the problem by court rule, and New York and California did so by common law decision in confused ways. Illinois correctly abolished the rule by condemning the erroneous and immoral rationale from the rule. Subsequent modifications should be admissible for all purposes. There should be little solace to those who complain that they would subject society to unreasonable risks but for the general rule.

Conformance to warning standards which were less than reasonable prudence. In real life there has yet to be a safety standard adopted that rises to the level of reasonable prudence. As legislatures rush to reinstate the unreasonable privilege eliminated by the common law there are all sorts of attempts by industry to excuse negligence by showing compliance with government warning standards which are totally inadequate. If the courts allow this ploy, the law is an ass.

This article has purposely not taken space either to cite or to quote from the reported decisions. They are very adequately reported in the *ATLA Law Journal,* the *ATLA Law Reporter,* the Appendix to the *Restatement of Torts 2d,* and by James B. Sales in an excellent rationalization for the backward law in the Defense Research Institute's *Monograph.* As lawyers on the side of people, we have only come a third of the way in righting the law of warnings. Sophistication in accident prevention needs and concepts and perseverance will get us the rest of the way.

Considerations in design

1) *Durability*—The warning must be on the product and be effective for the life of the product. If the machine is to be used for 100 years, then at the end of the 100 years the warning should exist and be as easily read as the manufacturer's name plate (so well designed because it is an advertisement for replacement parts).

2) *Signal word*—A signal word such as "Danger," "Caution," "Warning," or "Poison" classifies immediately the kind and character of the hazard and the gravity of the risk. For example, "Danger" indicates immediate and grave peril and by itself communicates that there is a hazard capable of producing irreversible injury. The signal word is the attention getter. The signal word for different risks is governed by the most grave risk.

3) *Color combination*—Warnings operate upon the unconscious mind as well as the conscious. Warning signs "speak a language of their own and operate together with the corresponding signal word to convey the extent of the peril in order to achieve virtually instantaneous recognition of the existence of the hazard and the gravity of the risk. Uniformity in design crosses language and literacy barriers in that it alerts the user of the necessity of getting further information before acting.

4) *Size of warning sign or tag*—The size of any warning and the panelization within the sign is an integral part of the exhortation to prevent injury and depends upon standards, viewing distance, illumination, length of message, etc.

5) *Size of type*—Type size should be large enough to permit easy reading at the viewing distance reasonably foreseeable. Emphasis may be achieved also by a limited number of capital letters, underlining or bold face or italic type.

6) *Message*—The message must detail the hazard—the gravity of the risk and the safety engineering alternative. It should be done with a strong active voice with short, simple sentences and active verbs. It should contain no surplusage. It should be in the number of languages reasonably foreseeable. It should depend in part on the time to view. *It should not be ambiguous.* Words and phrases necessary for communication should not be left out unless the designer is assured that an abbreviation or sign will be un-

. . . it is our social policy, as reflected in the law, which continues to be the problem.

derstood. Research indicates even well-designed symbols require some 50 practice trials. The designer can ill afford the luxury of 50 dry runs. The reason for precautions should not be buried. Inverted sentences are desirable.

7) *Sign placement*—Warnings must be placed to inform and exhort in sufficient time for appropriate action to be taken. They should be legible and not be competing with distractions. Potential blockage of the sign should be foreseen.

8) *Sign illumination*—Warning illumination is an important consideration and may require the warning to be of a particular color or require reflective material, emergency lighting, etc., since a foreseeable need for the warning may be during power failure, blackouts, fire, etc.

If one thinks of warnings as a vital component of accident prevention, then one is appalled at the disclaimers or liability avoidance statements on such things as diet pop cans where the risk of cancer is buried in verbiage, or football helmets where the tiny decal warning of paralysis, brain injury, or death is buried beneath the padding, or worse yet in the small print of the *Physician's Desk Reference* by the drug manufacturers. These schemes exist and flourish because the law has given and does give legal license to kill in that it is still so woefully underdeveloped in the protection of lives from injury.

Science and engineering in this area have not failed; it is our social policy, as reflected in the law, which continues to be the problem.

© 1981 Association of Trial Lawyers of America. Reprinted with permission from TRIAL, November, 1981.

References

Mr. Philo suggests the following additional reading for those interested in failure-to-warn issues.

1. NATIONAL MACHINE TOOL BUILDERS BULLETIN 69-4 (January 15, 1969).
2. American National Standards Institute Standards: Z35.1 (1972) Specifications for Accident Prevention Signs; Z35.2 (1968) Specifications for Accident Prevention Tags; Z35.4 (1973) Specifications for Information Signs; Z0.6 (1971) Manual on Uniform Traffic Control Devices for Streets and Highways; B14.1 (1971) Slow Moving Vehicle Identification Emblem (this is the same as ASAE S276.2 (1968) and S.A.E. J943); N2.3 (1967) Immediate Evacuation Signal For Use In Industrial Installations Where Radiation Exposure May Occur; Z53.1 (1971) Safety Color Code For Marking Physical Hazards; A13.1 (1956) Scheme For The Identification Of Piping Systems; and Z129.2 (1976) Precautionary Labeling Of Hazardous Industrial Chemicals.
3. ENCYCLOPEDIA OF CHEMICAL LABELING, Chemical Publishing Co., Inc. (1961).
4. SAFETY SIGNS FOR AGRICULTURAL, EARTHMOVING AND FORESTRY MACHINES. S.A.E. Publication 710707.
5. BULLETIN ON PRECAUTIONARY LABELS. American Petroleum Bulletin 2511.
6. GUIDE TO PRECAUTIONARY LABELING OF HAZARDOUS CHEMICALS. Manufacturing Chemists Association, Manual L-1 (1945).
7. Davis and Parswell. WARNINGS

AND HUMAN BEHAVIOR: IMPLICATIONS FOR DESIGN OF PRODUCT WARNINGS.
8. MODERN PACKAGING—A SPECIAL REPORT: HOW TO LABEL YOUR HAZARDOUS PRODUCT. A McGraw-Hill special publication (1968).
9. Man and Manager Inc. PRODUCT LIABILITY PORTFOLIO. Sections on Warnings and Recalls (1975).
10. *Labeling Hazardous Products.* National Safety Council paper delivered by Harry McIntyre at NATIONAL SAFETY CONGRESS (September 29, 1975).
11. PSYCHOLOGICALLY EFFECTIVE WARNINGS—HAZARD PREVENTION (May/June 1981).
12. DUTY TO WARN. Defense Research Institute Monograph No. 2 (1980).
13. Ross, Kenneth, *Legal and Practical Considerations For the Creation of Warning Labels and Instruction Books.* JOURNAL OF PRODUCT LIABILITY, Vol. 4, pp. 29-45 (1981).
14. JOURNAL OF THE ASSOCIATION OF TRIAL LAWYERS OF AMERICA, 36 ATLA L.J., 1-19 (1976).
15. *How John Deere Deals With Safety.* NATIONAL SAFETY NEWS (December 1970). *See also* JOHN DEERE DESIGN MANUAL #4, Safety, Instructional and Misc. Signs & Standard Letters and Numerals.
16. A GUIDE TO SIGNALLING REQUIREMENTS OF OSHA Federal Sign & Signal Corp. Monograph.
17. PRODUCT SAFETY AND LIABILITY LOSS CONTROL HANDBOOK. Alliance of American Insurers (1979). *See* pp. 15-21 on Labeling.

See also the catalogues of the warning sign suppliers listed below.
1. Legible Signs Inc., 2221 Nimitz Road, Rockford, IL 61111.
2. Stonehouse Signs Inc., P.O. Box 546, Arvada, CO 80001.
3. W. H. Brady Co., 727 West Glendale Avenue, Milwaukee, WI 53201.
4. Nutheme Illustrated Safety Co., 2020 Lunt Av., Elk Grove Village, IL 60007.
5. Emed Co., 1675 South Park Avenue, Buffalo, NY 14220.
6. Glas-Tex, P.O. Box 368, Baldwin Park, CA 91706.
7. Hazard Controls Inc., Woodward and Yale Avenues, Cherry Hill, NJ 08034.
8. Eastern Metal of Elmira, Inc., Industrial Park, 1430 Sullivan Street, Elmira, NY 14901.
9. Rockford Safety Equipment Co., 4620 Hydraulic Road, Rockford, IL 61109.
10. J. J. Keller & Associates, Inc., 145 W. Wisconsin Avenue, Neenah, WI 54956.
11. Inter-American Safety Council, 33 Park Place, Englewood, NJ 07631.

Fall accident patterns

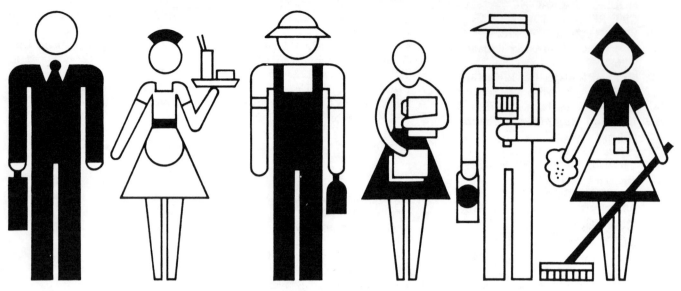

Characterization of most frequent work surface-related injuries

by H. Harvey Cohen and D. M. J. Compton

Accidents related to work surfaces are responsible for a large fraction of U.S. occupational injuries. One of the more complete tabulations, that for worker's compensation cases in New York State 1966-1970,[1] reported that 120,682 injuries with work surfaces as the agency of accident occurred during that period (20% of the total), with $219,152,000 awarded in worker's compensation (25.5% of the total).

Despite the magnitude of the problem, only a small amount of research on the causes and prevention of work surface-related injuries has been reported in technical literature.[2] Not only is there a clear lack of knowledge about the factors which contribute to work surface-related injuries, but there is also a deficiency in the documentation of practices and surface materials which may be effective for minimizing the risks of such injuries.

Therefore, the work reported here was directed, under contract to NIOSH,* towards developing accident profiles which describe the various industries, tasks, and work surface conditions characterizing the most frequent and serious work surface-related injuries. This was accomplished through two approaches. The first approach was the analysis of *existing* injury data. The second involved observations and measurements at 50 accident sites in 10 workplace establishments over a one-year period of time.

The sites for field measurements were selected to emphasize *slips and stumbles* that could be associated with coefficient of friction of the work surface/shoe combination, and, more particularly, slips and stumbles on *dry* rather than wet work surfaces. This was in order to study the feasibility of eventual incorporation of coefficient of friction requirements for dry floors in workplace regulations.

Analysis of injury data

Two types of injury data were analyzed: (1) tabulated data from the New York State worker's compensation agency, and (2) data from review of a sample of First Reports of Injury. Each of these sources was reviewed with the aim of: (1) determining the frequency of work surface-related accidents, especially those involving inappropriate coefficient of friction; and (2) identifying the basic characteristics and accident profiles of work surface-related accidents.

New York State worker's compensation data. Table 1 shows compensable injury rates (per 100 man-years) for New York industries showing a high risk of fall accidents. The New York publication previously cited lists number of compensable injuries by industry. These numbers were converted to rates using employment figures from County Business Patterns.[3] The New York publication also lists the number of injuries by accident type and by agency of accident for each industry. These are given as percentages of total injuries in Table 1 for two accident types (i.e., falls on same level and falls to a different level) and for work surface as the agency of accident.

The results show that the construction trades, e.g., painters, decorators, roofers, sheet metal workers, have a higher than average percentage of injuries as falls to different level. Office and service workers, restaurants, retail stores, hospitals, food and clothing manufacturers are examples of industry groups with a much higher than average fraction

*This work was supported by the National Institute for Occupational Safety and Health (NIOSH) under Contract 210-76-0150

PRIORITY INDUSTRIES WITH RESPECT TO WORK SURFACE AS AGENCY OF ACCIDENT

SIC No.	Industry Name	Total Injuries/ 100 man-years	Falls on same level	Injuries as % of total for SIC	
				Falls to different level	Work surface as agency of accident
15	General building contractors	0.7	8.3	19.9	29.8
172	Painting, paper hanging, decorating	12.1	6.0	38.2	47.6
173	Electrical work	2.1	8.9	19.5	31.5
174	Masonry, stonework, and plastering	5.6	8.2	18.9	29.8
175	Carpentering and flooring	5.2	7.0	22.4	31.3
176	Roofing and sheet metal work	5.6	4.8	26.7	32.8
177	Concrete work	6.8	10.9	16.0	28.2
1791	Structural steel erection	8.9	9.0	21.8	32.8
1799	Special trade contractors, NEC	4.3	6.6	23.3	32.4
202	Dairy products	7.2	14.9	11.7	25.4
231	Men's and boys' suits and coats	1.2	16.1	6.9	26.0
481	Telephone communication	0.6	12.3	12.1	31.9
531	Department stores	1.4	17.2	9.8	29.2
56	Apparel and accessory stores	0.7	20.6	17.2	41.3
58	Eating and drinking places	2.1	17.8	7.4	28.1
599	Retail stores, NEC	3.6	12.5	12.6	26.0
60	Banking	0.4	20.3	11.4	34.7
63	Insurance carriers	0.4	23.6	10.4	38.0
651	Real estate operators and lessors	2.1	13.3	17.5	31.5
701	Hotels, tourist courts, and motels	2.4	19.6	10.3	32.1
734	Services to buildings	2.3	15.9	14.2	31.7
739	Miscellaneous business services	0.9	16.5	12.3	29.8
80	Medical and other health services, excluding hospitals	0.8	21.3	8.7	32.0
806	Hospitals	0.9	16.7	5.5	24.9
81	Legal services	1.0	20.4	12.1	35.6
864	Civic and social associations	3.0	15.6	15.6	28.5
867	Charitable organizations	2.0	22.1	12.1	35.5

Table 1

of their compensable injuries being falls on same level.

First reports of injury. In order to gain more insight into the accident profiles characteristic of slip and fall accidents, a series of approximately 3,000 First Reports of Injury were analyzed. These reports were obtained during a previous NIOSH study[4] during which 621 establishments were visited nationwide, and copies of First Reports of Injury were collected for a one-year period. In all, about 22,000 First Reports were available. The selection of establishments in this prior study was designed to include the 25 two-digit SIC codes with the highest number of injuries (i.e., highest product of injury rate and employment), as well as a range of size and geographical distribution.

These 22,000 First Reports were reviewed and all injuries involving slips or falls on a work surface were selected for further analysis. The total number of injury reports so selected was 3,270 (approximately 15% of the total).

Table 2 shows the distribution of accidents by the event that immediately led to the injury. The majority, or about 50%, of the cases were characterized as "slips." These events resulted in the accident types (results) shown in Table 3. Notice that at least 25% of the events resulted in *incomplete* falls.

Table 4 shows the distribution by occupational category. Although unskilled laborers make up only 12% of the work force, they experienced about 27% of the injuries. Other occupational groups which show a high incidence of fall injuries include various types of skilled, office, and other "white collar" workers.

The type of work surface involved is shown in Table 5. Not surprisingly, the most frequent surface indicated was the "floor." This is probably more a function of employee exposure than relative risk potential. Other surfaces frequently involved were: ground (outdoors), stairs, and ladders.

The other interesting findings of

EVENTS THAT LED TO THE INJURY	
Events	% of Total Cases
Slip	50
Trip	14
Misstep	10
Loss of support	7
Postural overextension	4
Stumble	1
External force	1
Medical incident	<1
Unknown	12
	100

Table 2

DISTRIBUTION BY ACCIDENT TYPE (RESULT)	
Broad Accident Types: Results	% of Total Cases
Fall on same level	26
Fall to a different level	19
Recovery on same level	14
Struck against object during fall: *incomplete fall*	10
Recovery to a different level	1
Fall unknown	10
Recovery unknown	1
Other	<19
	100

Table 3

DISTRIBUTION BY OCCUPATIONAL CATEGORY	
Occupational Category	% of Total Cases
Unskilled laborer	27
Skilled laborer	19
Clerical	8
Transport equipment operator	8
Machine operator	7
Production worker	7
Manager/supervisor	6
Maintenance worker	5
Service worker	5
Trainee	1
Other occupation	2
Unknown occupation	<5
	100

Table 4

DISTRIBUTION BY TYPE OF WORK SURFACE	
Work Surface Type	% of Total Cases
Floor	33
Ground (outdoors)	11
Stair	10
Ladder	6
Ramp/slope/incline	2
Scaffold/catwalk	2
Other surface	21
Not stated	<15
	100

Table 5

the injury report analysis are:
1. About two-thirds of the reported accidents occurred indoors, while fully one-third occurred out-of-doors.
2. Over 11% of the cases were reported to occur at "temporary" rather than "regular" work sites.
3. Of the slips, 16% occurred on surfaces stated to be wet, 8.1% icy, 6.4% oily, and 0.7% muddy. An additional 3.5% of the surfaces were characterized as "slippery."
4. At least 5% of the cases involved pulling motions and 3% involved pushing forces.
5. At least 4% involved a leaning position, while 2% involved postural overextension.
6. In 13% of the cases, the injured worker was carrying an object.
7. The most frequent foot motions were walking, 32%; stepping onto/from (such as a ladder), 21%; standing, 15%; and climbing, 3%.
8. Twenty-two percent of the cases indicated poor or inadequate housekeeping.
9. Twenty-one percent involved poor or inadequate lighting.
10. Twenty-seven percent implicated haste as a factor.

Site observations

In order to further characterize work surfaces and tasks involved in fall accidents, 50 site observations were made of locations in which recent fall accidents had occurred.

Selection of sites. Selection of establishments in which to make site measurements was based on the following criteria: (1) industrial classification, i.e., those having a high incidence rate of slips and falls, as shown in Table 1; (2) a reasonable variety of sites; (3) ability to compare two establishments in the same industry to determine possible differences in practices and injury experience; and (4) proximity to the study team.

Nine organizations in San Diego County, California, agreed to participate in the study, make available their injury data, and allow access to sites for observations. One organization (a university) also operated a hospital, which was separately identified in this study, giving 10 establishments in all, as listed in Table 6.

Initial familiarization visits were made to the participating establishments. First Reports of Injury, representing a period of approximately one year of experience, were reviewed in order to determine the extent to which the injuries were related to work surfaces and to make a preliminary selection of sites for possible observation and evaluation. Actual selection of sites was based on relevance to the project and feasibility of data collection.

For all the establishments taken together, only 35 slips and two stumbles occurred on dry surfaces in the one-year period. These represented only about 5% of their total slips and falls. The 50 site observations were, therefore, selected to include all of the 35 sites at which "dry" slips had occurred, the two sites at which stumbles had occurred, and 15 sites at which slips had occurred on a "wet" surface.

Conduct of site observations. An attempt was made in each site evaluation to acquire the most complete and detailed data possible regarding the condition and use of the site, the circumstances of the accident, and the tasks and body positions of the injured employee. After all available data had been gathered, a visit was made to the site to observe and document the physical parameters related to each event. The following items were covered:

1. *Footwear parameters.* Shoe type, fastening type, sole and heel height, and proper fit were evaluated. Sole and heel construction material and condition (extent of wear) were identified so as to evaluate the role of footwear in the coefficient of friction at the site.
2. *Work surface characteristics.* Physical parameters of work surfaces which were routinely observed were: (1) surface type and function [hall, walkway, stair, plant floor, scaffold, entrances, etc.]; (2) surface construction [materials used, texture, and visibility; design features such as incline, drainage, unusual topographical features]; (3) surface treatment [acid-etched, waxed and buf-

DESCRIPTION OF PARTICIPANT ESTABLISHMENTS

Code No.	SIC	Approximate number of employees	Fall incidence rate	Operational description
601	9199	9,000	0.38	Local government
602	9199	6,000	3.71	Local government
603	3731	6,160	2.71	Ship construction, repair, conversion
604*	8062	1,885	1.23	Medical and surgical hospital
605	8062	919	2.43	Medical and surgical hospital
606	37	6,000	2.03	Transportation vehicle manufacturer
607	8221	2,700	1.33	College, university
608	8221	5,000	0.76	College, university
609	3661	500	3.68	Telecommunications apparatus manufacturer
610	2900	2,000	1.99	Fast food restaurant chain

*Operated by the same organization as Establishment No. 608.

Table 6

fed, nonskid material, etc.]; (4) surface condition [maintenance features such as wear, scuffing, ill-repair; hazardous condition features such as water, oil, ice, dust, etc.]; (5) layout [adequacy of work space, traffic routes, work stations, adjoinments of two surfaces with widely different properties related to coefficient of friction]; and (6) subjective estimates of slipperiness.

3. *Task parameters.* Observation of the types of jobs performed in and about the site was made to determine relevancy to the accident occurrence. Generic types of tasks such as transit, materials handling, office work, etc., were identified. How these tasks were performed—processes, procedures, and equipment used—was also evaluated. Traffic volume and traffic patterns of persons working in the immediate area from other areas were observed. Areas of shoe sole contact with the work surface as associated with the performance of specific tasks were investigated. An attempt to understand the forces and weight distributions imposed on the body was made.

4. *Housekeeping parameters.* Some housekeeping issues relevant to work surface-related injuries are: (1) establishment operations (such as leaky machines) and procedures which contribute to housekeeping problems; (2) water, oil, etc., spills, and policies for reporting them; (3) types of surface finishing treatments used and the criteria, if any, used for selection; (4) wax stripping, waxing, and buffing schedules and whether workdays were allowed to elapse between operations; (5) frequency of dust-mopping and sweeping operations; (6) loose objects on the floor; and (7) isolating in-progress stripping, waxing, and mopping areas from foot traffic.

5. *Environmental parameters.* (1) lighting type and adequacy; (2) heat and humidity and their possible effects on the properties of wax, other floor surface treatments, and on shoe soles; (3) ambient weather conditions such as precipitation and fog; and (4) noise and other distracting factors.

6. *Work surface COF parameters.* Coefficient of friction (COF) measurements were made at a number of locations depending on the need to understand the circumstances of each accident. These included: (1) the accident site; (2) approaches, where applicable, to the accident site; and (3) locations where the injured employee normally worked if it appeared that an unexpected change in COF may have been involved in the accident circumstances. At least five measurements of COF were made at each site. Mean ($\bar{x}$) values and standard deviation(s) were computed. T-test statistics were used to test for the significance of differences between two sets of COF observations.

A number of instruments using a variety of principles have been developed for measuring COF[2]. In this study, four instruments were evaluated and intercompared.

1. *The Universal Friction Testing Machine (UFTM)* was a motor-driven rotary-motion device supplied by NIOSH for purposes of feasibility testing. It rotated two 25/31-inch-diameter footwear sole samples under a constant vertical pressure and provided both static and dynamic COF measurement capability. A range of speeds and a digital readout of COF was also provided.

2. *The Olson Horizontal Full Slip Meter* was a commercially available device which employed a constant-torque motor to drag a weighted scale with three ½-inch pads mounted with footwear sole samples across the test surface so as to measure both static and dynamic COF.

3. *NBS Brungraber Portable Slip-Resistance Tester.* This nonpowered device was developed and supplied by the National Bureau of Standards (NBS). It uses the principle of the articulated strut but measures only static COF. A weight is attached to a shaft articulated at an angle approximately equal to a COF of 0.03. The angle of articulation increases until the sole sample slips. The tangent of the angle between the articulated shaft and vertical is related to the COF.

4. *BIGFOOT.* Because of a number of field operational difficulties encountered during trial tests of the above instruments, a simple manually operated horizontal sliding device

was developed "in-house," called BIGFOOT. It consisted of a footwear sole sample holding bracket which allowed for the mounting of a 3½ by 4¾ inch shoe sole sample. This large format minimized susceptibility to variations in readings due to irregular surface topography. A 10-pound weight was mounted on top of the bracket and the whole unit was pulled horizontally by a 1-10 pound spring balance (Chatillon gauge—R Cat 719-10) equipped with a peak-reading device which gives the static COF measurement. Dynamic COF could also be measured and read by pulling the unit at a constant velocity and observing the scale readings. The weight could also be removed from the bracket, placed in a shoe and with a cord, pulled across the test surface. This enables measurements to be made with the actual shoe involved in an accident.

Intercomparison of the four measuring instruments. The four COF measurement devices were intercompared in order to identify which device would yield the most reproducible data from the least number of measurements.

A test program was devised which called for a uniform surface, a clean Formica laboratory table top, leather footwear sole samples sanded at regular, consistent intervals, and 100 static measurements made with each tester performed by one operator. The resulting measurements were then analyzed. Table 7 shows the means and standard deviations from this analysis.

The mean of the mean COF values obtained was 0.345, which was used as a central midpoint to plot histograms for each of the four distributions of data.

Analysis of the standard deviations indicated that, despite the potential for low operator reliability due to variability in manual pulling speeds and pulling techniques, the BIGFOOT device had the smallest standard deviation and, therefore, would be expected to provide the most reproducible data. Further, its mean COF, 0.335, lay closest to the mean of all the mean COFs, 0.345.

Intercomparison of different sole materials. A series of measurements were made to intercompare COF with different sole materials and to compare the results with subjective impressions of the slipperiness of various surfaces. Figure 1 shows COF measured using six sole materials on six surfaces, selected to have widely different subjective degrees of slipperiness. The values for each surface are plotted against the average of all the COF measurements for that surface. The six sole materials were: (1) leather; (2) Biltrite, a cork composition; (3) Chemgum, a rubber-like material; (4) Neoprene; (5) Avalon Thrust, an expanded material; and (6) Hypalon Quabaug, a reinforced rubber material. The six surfaces were (1) terrazzo tile with brass edging; (2) ceramic floor tile; (3) unwaxed vinyl tile; (4) vinyl sheet, waxed and buffed; (5) concrete, walkway standard; and (6) concrete, rough walkway standard. The above surfaces are in order of decreasing slipperiness as perceived by a number of observers.

Examination of Figure 1 shows that the measured COF generally decreased as subjective slipperiness of the surface increased. The sole materials also show generally consistent COF behavior as the surface is varied. In this sense, some surfaces and some sole materials can be described loosely as "high COF." An interesting observation is that the high-COF sole materials appear to show a more rapid dropoff in COF as the surface becomes more slippery than is shown by the lower COF materials. A line is drawn in Figure 1 to indicate the range considered to be "slippery."

It should be noted that slipperiness may be related not only to the absolute value of the COF but to the difference in COF "felt" by a worker between one location and another. If the worker does not adjust his gait, a slip may occur even with a high absolute COF value, while a worker accustomed to a low shoe-floor COF may be able to avoid slips.

Floors were considered potentially slippery for COF readings below the following values: (1) leather, 0.42; (2) Biltrite, 0.46; (3) Chemgum, 0.60; (4) Neoprene, 0.55; (5) Avalon Thrust, 0.70; and (6) Hypalon Quabaug, 0.78.

Measurements on wet surfaces. Wet spots on a floor such as vinyl tile or ceramic tile are well known to be more "slippery" than the dry floor, and are a factor involved in many slips and falls. It would be very desirable to be able to measure the degree to which water makes a flooring material more slippery. Unfortunately, most methods of COF measurement give inconsistent results on wet surfaces and often give results that are not in agreement with subjective experience. In many cases, the measured COF *increases* when a surface is wetted, even though the surface is known to be slippery when wet. It was found that, if measurements were made using water with a wetting agent such as a household liquid detergent, results more consistent with subjective experience were obtained.

Summary of site observations
Despite the eight types of fall accidents indicated in Table 1, only slips and possibly stumbles are potentially related to COF. Slips often are associated with low-COF conditions, while stumbles may result from high-COF situations, e.g., where the foot is "caught" on the surface.

In general, slips occur when the parallel force, Fp, exceeds the product of the COF and the normal force, Fn. For a constant COF, this situation is observed to arise under the following circumstances:

1. *High horizontal forces.* When the worker is pushing, pulling, accelerating in walking speed (including and especially when turning a corner or sidestepping), jumping, throwing, or catching an object.
2. *Lowered vertical forces.* These occur when the worker "bobs down," i.e., rapidly bends the knees to unweight the feet, or if the work surface gives way beneath the worker's foot.

INTERCOMPARISON OF MEASURING INSTRUMENTS			
Device	Mean	Standard deviation	Standard deviation/ mean
UFTM	0.31	0.064	0.206
BRUNGRABER	0.36	0.039	0.108
OLSON	0.378	0.032	0.085
BIGFOOT	0.335	0.023	0.069

Table 7

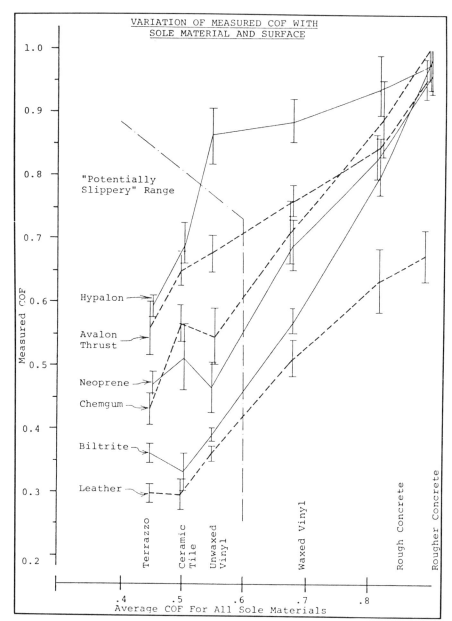

Figure 1

However, once a slip starts, the COF will have its dynamic (lower) value, rather than its static value. In addition, the angle between the leg and vertical is likely to be increasing rather than decreasing, and additional parallel forces will be needed to decelerate the moving foot to rest. Physical termination of a slip is thus unlikely except for very small low-COF patches. The maximum size of such patches is not currently known. Conversion from static to dynamic COF conditions may also occur in pivoting on a foot, where most of the shoe surface is made to move relative to the work surface.

Stumbles are often related to an unexpectedly high COF, so that a foot is "caught." The circumstances under which this was observed to arise include:

1. *Climbing stairs.* Many persons, while climbing stairs, place their feet on the steps with a short controlled sliding motion. If the COF is unexpectedly high, control may be lost, with the foot "caught" and delayed, typically followed by a trip over the next step. Similarly, when descending stairs, a person may stumble when either the sole or heel of the trailing shoe catches.

2. *Transition from smooth to rough surface.* Many persons were observed to stumble when walking from a smooth onto a rougher surface such as a carpet. This may be due to a controlled glide type of walking on the smooth surface, in which the feet are elevated much less than usual and allowed to slip loosely over the surface during the forward swing. If the same gait is attempted on a rough surface, the feet "catch" and a loss of control results.

Controls top priority

Analysis of injury reports, supplemented by detailed site observations, reveal that fall accidents are likely to be most frequent and severe when certain combinations of activities, conditions, and circumstances occur, as follows:

1. While passing through *doorways,* where there is often a change in surface, level, and lighting, wet carried indoors from outdoors, a threshold strip of different COF, a need

Similarly, a worker who steps, unaware, onto a surface that is lower than the rest of his path has lower than normal vertical forces.

3. Angle between leg and a horizontal work surface. As described above, placing the leg at a large angle to the vertical results in increased parallel direction of body weight.

4. Angle between horizontal and work surface. If the work surface is not horizontal, the vertical forces associated with the body or other weight will not be normal to the surface; vertical forces will have substantial components parallel to the work surface. This may arise when the shoe is placed on a rounded object, and the ankle cannot keep the foot level. The shoe then tilts, so that there is an angle between the horizontal and the normal to the surface, and the shoe "slides off" the object.

Combinations of the above circumstances were observed to arise in the case of a worker trying to push a cart up an inclined ramp.

Recovery from a slip depends on the recovery of balance or on the termination of slipping for physical reasons, or both. Recovery of balance from a slip (i.e., an incomplete fall) depends on the dynamic postural reflexes, through moving the center of gravity relative to the point of support (e.g., through flinging out an arm). This may involve injury through muscle strain. Slipping may also terminate for physical reasons, e.g., if there is a small patch with low COF on a high-COF work surface.

to push or pull the door, and a traffic funnel where conflict requires sidestepping or other evasive action.
2. While working on *loading docks,* where there is an unguarded edge, a need for high horizontal forces in pulling, pushing, and throwing objects, and a work surface that may be wet from rain or oily from lift trucks.
3. While *dismounting* (in a backward direction) from a ladder, vehicle, scaffold, or work platform. The surface onto which the person dismounts requires freedom from oil, spills, loose objects, equipment, holes, depressions, etc.
4. When *handling hoses,* which tend to leak water or fuel, resulting in slips, or *other long coiled objects,* such as rope, wire, welding cable, which require pulling motions, typically in a backward direction, thus obscuring vision, and resulting in a tripping or misstepping hazard.
5. While *carrying heavy or bulky objects.*
6. When encountering *unexpected work surface hazards* (both indoors and outdoors) in unfamiliar work areas.
7. In situations where there is *insufficient workspace* around level work areas or surrounding elevated work stations.
8. Where there is a *lack of communication or coordination* between coworkers, e.g., a hose or welding cable wrapped around a ladder.
9. While using *makeshift platforms* to extend one's effective reach.
10. When *overreaching* laterally, extending beyond the base of support, standing above the top three rungs of a ladder, setting up a ladder on an unstable support, or transferring laterally from a ladder to a platform.

The field portion of the study focused on COF as a contributory factor to slip and fall injuries. It should be noted that slip and fall injuries on dry surfaces were found to be rather rare occurrences, accounting for about 5% of the establishments' total slips and falls, or about 1% of their total injuries.

Although the number of injuries of this type is small, the site measurements indicate that control is possible. This is based on the following findings made during the site observations:

- COF varied from below 0.2 to nearly 1.0 over the range of work surfaces tested.
- In one plant with a COF of 0.2 (extremely slippery) on a specially treated concrete floor, employees were able to adjust their gait to walking but tended to fall when performing tasks such as reaching and pushing.
- Measurements over a floor in one large building showed that the main corridor had a consistent COF of about 0.4 (adequate). The rooms leading off the corridor tended to have COF values of about 0.6, and some side corridors were slippery with a COF of 0.2 to 0.3. Such variation in COF places a burden on unsuspecting employees.
- Employees generally seem to be able to adjust to a wide range of COF, *provided that it remains uniform* and they are not overloaded by task demands.
- *Unrecognized changes of COF* appear to account for most slips on dry surfaces, typically on walking from a high-COF surface to a lower-COF surface without clear visual cues.
- The same factor of *lack of recognition* appears to apply to wet and oily spots on surfaces as a factor involved in slips.

Control of COF by etching of tile and concrete floors, and by proper waxing and buffing of vinyl floors, is quite feasible and is performed systematically by some of the organizations studied. For example, a hospital with a thorough floor care program, which includes COF measurements, has a rate for slips and falls of only 28% of that of a nearby hospital with a less effective floor care program. Field measurements confirmed a significant difference in COF between a series of measurements in one hospital as compared with the other. This suggests that *adequate floor care* would be an effective countermeasure for many slips of this kind.

Other control measures shown likely to be effective include: (1) use of nonslip work surface materials; (2) careful inspection and maintenance of key trouble spots, and (3) improved housekeeping.

Several factors require additional research. For example, the situation regarding measurement of COF is unsatisfactory. There is no generally accepted instrument or procedure and no standard samples of flooring and footwear that are readily available for field calibration purposes over a range of COF. The situation is worse for measurements on wet surfaces, where most slips occur. The results of the study did indicate, however, that use of a wetting agent (i.e., a common household, liquid detergent) appears to improve the usefulness of COF measurements on wet surfaces.

Shoe design is an obvious priority for future research. The materials that have a high COF on "high-COF" floor surfaces tend to give a rapidly decreasing COF as the flooring increases in slipperiness. Also, there are some situations which cannot easily be otherwise controlled, e.g., a welder in a shipyard who pulls on welding cables and stands on wet steel covered with welding grit.

It does not appear that effective workplace criteria regarding COF can be promulgated at this time. However, immediate attention can be focused on training needs associated with specific accident profiles for occupational groups in especially high risk of falls.

References

1. "Characteristics and Costs of Work Injuries in New York State, 1966-1970. Volume 1," State of New York, Department of Labor, Division of Research and Statistics. *Special Bulletin No. 243.* December 1972.
2. Pfauth, M. J. and J. M. Miller, "Work Surface Friction Coefficients: A Survey of Relevant Factors and Measurement Methodology." *Journal of Safety Research,* 8:77-90, 1976.
3. County Business Patterns, Bureau of Census, 1973.
4. SAFETY SCIENCES, "Feasibility of Securing Research-Defining Accident Statistics," *DHEW (NIOSH) Publication No. 78-180,* September 1978.

Entering and exiting elevated vehicles

by Ron Hurst and Tarek Khalil

There have been many reported cases of operators falling and injuring themselves while exiting elevated vehicles such as trucks, tractors, trailers and crane cabs.

This paper ergonomically analyzes various examples of means of entering and exiting elevated vehicles. Biomechanical analysis with respect to stability, space, and balance is provided. The paper reviews problems of design and causes of injuries. Appropriate standards are cited and the approach to recognize, evaluate and control this type of safety and health hazard is illustrated.

Accidents involving "slips and falls" on floors and stairways are a major cause of injuries throughout the United States. One problem area that falls under this category is "slips and falls" in and around powered vehicular equipment, such as cranes, forklifts, sit-down rider trucks, loaders, pavers, and rollers. One of the major contributing tasks in which these accidents occur is the access/exit or mounting/dismounting of the vehicle. This is the task of getting into and out of the cab itself just for the purposes of driving or performing a major task connected with the vehicle.

It has been estimated by Miller[1] (1976) that about one-fourth of all driver injuries are associated with slips and falls in and around the truck, and these injuries may be responsible for a large percentage of the permanent, long-term disabilities, particularly with respect to back injuries. It should be pointed out that back injuries result in about one-third of all occupational injuries and one-third of all compensation associated with injuries.

While falls from vehicles are not among the most frequent accidents, according to McPeak[2] (1976), they rank second in number of indemnity days per case, and third in average amount of workers' compensation benefits paid. This kind of accident ranks relatively low in mean number of healing days, and the leading part of the body injured was the lower extremities along with the lower back. McPeak concluded that these facts would indicate that a great many of the falls were not from a great height.

A recent study by Sparrell[3] (1980) found that out of 47 severe or fatal injuries involving off-highway vehicles, six cases involved operators who suffered permanent partial disability as a result of falls as they dismounted from equipment. The severity of these injuries is indicated by medical and compensation payments in excess of $20,000 in each case. Sparrell also noted that he found in other studies that falls while mounting or dismounting are the most frequently reported source of injury occurring to equipment operators.

A study of Laumann (1977) indicates that notable numbers of injurious falls from trucks can be traced in part to poorly designed means of mounting and dismounting the vehicles.[4]

An earlier study by Hall[5] (1967) reported similar problems of accidents occurring while operators mount or dismount industrial equipment. He suggested that improvements in the vehicle design should be made. In spite of awareness of this problem by safety professionals for many years, we still find poor access/exit system designs today, and the accidents still occur.

Our own experience indicates that a significant number of product liability lawsuits exist as a result of injuries occurring during mounting/dismounting of vehicles.

Main problem areas

Based on a thorough review and analysis of a large number of vehicles actually in use, the following classification of the main problem areas were made:
1) hidden steps not permitting visual contact by operator
2) lack of uniformity in design of steps and vehicles
3) excessive height of step risers
4) lack of or poorly placed side steps
5) lack of or mislocation of handrails
6) other miscellaneous design problems
7) poor tread surface conditions
8) environmental conditions
9) lack of enforceable standards
10) lack of training

Hidden steps

Many current vehicle designs employ flat vertical sides. This design is readily suitable for building in good steps up to the driver's platform (Figure 1). Laumann pointed out this problem in 1977. He noted further that although steps could be cut into such vehicle sides, they would not be visible from the driver's seat. The operator would have to search around for the step while dismounting. Lack of visual contact could cause the operator to miss the step, thus leading to a slip and/or a fall.

The solution to reduce the problem of hidden steps lies within the design of the vehicle. The designer must take into consideration the means of access/exit before he designs the vehicle. If this consideration is taken into account, there would not be a need for hidden cut-in steps. The designer should provide an access or a ladder with easily visible steps leading from the driver's compartment to the ground.

Figure 1. Cut-in steps are hidden from driver's compartment and force driver to search for step.

Lack of Uniformity

The problem exists in two forms:
a) lack of uniformity from one step to the next, and
b) lack of uniformity from one vehicle to another

The lack of uniformity poses a problem because the user has learned to expect and depend upon a uniform stair construction. The user's perception and expectation are especially important. Successful stair usage is based upon the user's matching of internalized images and expectations to the actual structure which is present.[6]

An accident could be caused by a human perceptual error which is triggered by irregularity or non-uniformity in the design or construction of the access/exit system. When a perceptual error occurs, the user often cannot adjust behavior and motor responses in time to compensate for the error.

The solution to this problem would be to design an access or ladder with uniform steps leading from the ground to the operator's compartment.

In industry, an operator may be called upon to operate various vehicles, and may be exposed to many different designs of access/exit systems. The operator may have learned certain patterns of mounting and dismounting from one particular vehicle. When using a different vehicle with a different design, the operator may make a perceptual error which could result in a slip or fall.

A possible solution would be that, as far as possible, the industry should have a standard type of access/exit system on all vehicles that are used for similar tasks.

Step riser

The most common and most serious problem is that the first step from the ground is usually too high.

The Society of Automotive Engineers[7] recommend in SAE J 185 that the maximum height of the first step from the ground should not exceed 30 inches. The preferred height of the step is 16 inches. The maximum distance between each succeeding step of a vertical ladder on a vehicle should be 16 inches, but the preferred distance between steps is 12 inches.

On the other hand, OSHA[8,9] requires in Sections 1910.27(b)(1)ii and 1910.179(c)(2) of the 29 CFR 1910.10 "General Industry" Standards, and in Section 1926.450 of the 29 CFR 1926 "Construction Industry" Standards that the distance of the first step and each succeeding step shall not exceed 12 inches, and shall be uniform throughout. ANSI 14.3 "Safety Requirements for Fixed Ladders"[10] also requires that step

Figure 2. A vehicle where the height of access is 45 inches above ground.

heights shall not exceed 12 inches. Figure 2 shows a vehicle where the height of the first and only step to the operator's cab is 45 inches with no other means of mounting or dismounting.

A biomechanical analysis of steps indicates that there are three specific problems with riser height. The problems are a) If the step riser is too high, the user can only make point contact with the step instead of whole-foot contact; b) The center of gravity of the user will tend to shift outside the base of support affecting stability, and c) The user is subject to a rearward moment of force.

Figures 3 and 4 provide diagrammatic illustration of this problem. In the posture shown in Figure 3, the center of gravity (w) falls within the base of support of the body bounded by the distance between the two feet. The user can balance without excessively stressing the musculoskeletal system. In Figure 4, where the step is too high, the center of gravity (w) falls out of bounds of the base of support forcing the user's body into an unstable and off-balance position.

In Figure 4, the center of gravity (w) falls rearward of the supporting foot (F_1). (F_2), the pushing force on the step, causes a user's momentum in the rearward direction; thus, pushing the user away from the vehicle.

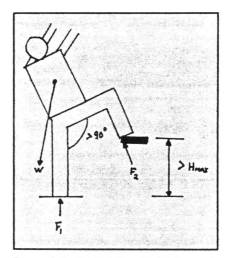

Figure 4. Individual is out of balance, is making point contact with the step and there is a rearward moment of force pushing user.

SAE Standard J 185 recommends that the user maintain three limbs in contact with the vehicle or ground while he or she is mounting or dismounting—the three-point system principle. The three-point system is a sound engineering principle for stability. However, many existing access/exit system designs do not provide this stability.

One example of when the principle would fail is when the last step to the ground is too high for the user during dismounting.

A possible solution to the problems associated with the step riser height being too high is to design a system that has steps (including the first step from the ground) that are of heights compatible with the physical limitations of the human population. The proper height allows the user to lift his or her leg to a height where the leg is bent with a 90° angle at the thigh with reference to the frontal plane while maintaining full foot contact with the step. (Figure 3).

This height can be calculated by subtracting the popliteal height (the height of the underside of the upper leg above the footrest surface, Figure 5) of an individual from the gluteal furrow height (the height of the furrow where the gluteal curve intersects the back of the thigh, Figure 6).[11]

$H(max) = GF - P$ (1) where:

$Hmax$ = Maximum allowable height of step (in)
GF = Gluteal Furrow Height (in)
P = Popliteal Height (in)

To design an access/exit system for the user population, we would suggest using the anthropometric data[11] compatible with 90% of the possible user population. We find that the popliteal height of women is 15.0 inches while the gluteal furrow height is 26.2 inches. If we plug these measurements into equation, we find that the recommended height is 11.2 inches.

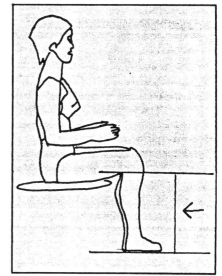

Figure 5. Popliteal Height

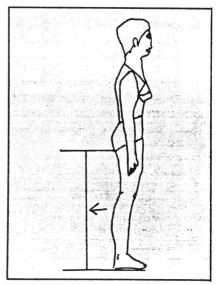

Figure 6. Gluteal Furrow Height

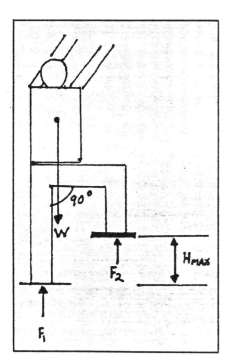

Figure 3. Individual is in balance and is making whole-foot contact.

One may argue that the operators of industrial equipment are much larger than that of the normal population, therefore, there is no need to include smaller individuals in the design requirements. This argument is not valid from a human engineering point of view. With the increasing number of women entering the work force, equipment should be designed to accommodate both males and females.

It is evident from this analysis that OSHA and ANSI standards provide more appropriate guidelines than those of the Society of Automotive Engineers. It would be prudent for a vehicle manufacturer to take a conservative approach in designing access systems in order to avoid product liability problems.

Side steps

OSHA 1910.27(d)(2)(i) states that where an individual has to step a distance greater than 12 inches from the centerline of a ladder to the nearest edge of equipment, a platform shall be provided. ANSI 14.3, Section 4.4.3.2 states that side-step ladders at the point of access/exit to a platform shall have a step-across distance of 15 inches minimum and 20 inches maximum from the centerline of the ladder.

Many vehicles violate these standards. Most vehicles lack a platform for the user to sidestep from the driver's compartment to the exit system. The problem arises when the user must reach from the ladder to the driver's compartment. In doing so, the user is subject to a moment of force pushing him in the direction of the reaching leg, therefore, creating a drastic shift in the center of gravity. The result is a possible loss of balance, missing of a step, or the forcing of the user to grab an inadequate handhold.

Possible solutions are to install the platform or to reduce the side-stepping distance. This would enable the user to maintain stable and balanced forces and to maintain a vertical center of gravity, which would reduce the risk of falling.

Handrails

There are four problems associated with the design of handrails of existing access/exit systems. They are lack of handrails, only one available handrail, mislocation of handrails, and an unstable point of contact.

Figure 7. Vehicle has only one handrail which forces the user to grasp an unstable handhold, the swinging door.

Lack of handrails. Handrails are used in the climbing gait as vehicles of gripping, pulling, and support by the user. If the user is to follow the recommended three-point system of mounting and dismounting, he must be able to have a point contact for his hands.

One handrail. If a vehicle has only one handrail (Figure 7), the user may not be able to follow the natural "climb pattern," depending on one hand and the opposite leg to make the upward move. When the user raises his leg opposite the hand lacking the handrail, the user must find another surface for handhold support. If this surface is inadequate or unstable, the user would lose his balance. Another problem is that a single handhold would cause the body to swing around the gripping point. This would cause a constant shift in the center of gravity of the individual which may be difficult to compensate for.

Mislocation of handrails. Handholds must be placed to follow the natural "climb pattern" of the user. They should be strategically located to match the user's path of ingress or egress.

Unstable handhold. We found that in many cases the user is forced to use a steering wheel or a swinging door as a handhold support. As we look at Figure 8, we see the effect of an unstable handhold. The individual is climbing steps with one hand gripping a stable hold and the other gripping the door. As the door swings open to the left, there is a moment of force rotating the individual to the left, potentially causing a fall from the vehicle.

A possible solution to these four problems would be to design a con-

Figure 8. The user is gripping an unstable source (the door) as a handhold. The swinging of the door causes a rotating force on the individual that leads to a shift out of balance.

tinuous handrail on each side of the ladder, and to have the handrails continue with any side step. The two continuous handrails will eliminate the user searching for handgrips, will meet the user's path of mounting and dismounting sequential hand position requirements, will meet the requirements of the natural climb pattern, will maintain the user's center of gravity, will eliminate the use of an unstable point of contact, and will allow the use of the three-point method of mounting and dismounting.

Miscellaneous design problems

Several other problems in the access/exit systems were found that did not fit into the classification system used here. It was felt that they were dangerous enough to be worthy of discussion.

In Figure 9, we find a vehicle that has an unstable step that hangs from link chains. When the operator puts his or her foot on the step, the step swings causing the operator to swing also. The operator is, therefore, at a high risk of losing balance and falling. The solution to this problem would be to ensure that all steps in the access/exit system are stable and solidly fastened.

Some vehicles simply do not have an access system. This lack of access/exit system is hazardous and puts the operator in a high risk area,

Figure 10. Good tread design. Ridged edges increase traction; drainage holes drain excess water, ice, mud, snow, etc.

Figure 9. Vehicle has an unstable step that hangs from a link chain and swings when it is stepped on.

particularly, if he or she has to abandon the vehicle in case of emergency.

Some vehicle designs invite the operator to perform unsafe acts; for example, the operator finds it convenient to climb over the shovel of the bulldozer in order to mount or dismount the steps of the access/exit system. This is extremely dangerous, especially in a situation where the operator may miss the step or lose his or her balance, thus, falling and striking the shovel.

The solution to this situation would be to make the access routes clear, unobstructed and nonsubstitutable.

These represent just a few miscellaneous hazards. The designer should make a concerted effort to recognize all potential hazards and make every practicable effort to avoid them.

Tread surfaces

Tread effects on contact area are very important because if frictional contact is lost between the user's shoe and the step, full slippage may result. A skid-free surface is important. To insure good friction and reduce the possibility of slipping, the operator should be required to wear work shoes that have a high coefficient of friction, such as rubber-soled shoes. The tread should also be of a material with a high coefficient of friction, such as a rubber mat, or the tread should be designed in a manner, such as Figure 10, where the ridged grill increases traction.

Environmental conditions

The presence of foreign substances such as grease, oil, dirt, water, snow, or ice on the steps could cause the user to lose frictional contact with the tread which can result in slippage.

We would like to make three suggestions as possible solutions to the problems associated with the tread which can result in slippage.

a) To design the step with holes, such as in Figure 10, which will drain excess substances from the step.
b) To establish a program in which the operator of the vehicle is responsible for inspecting and maintaining the access/exit system at the beginning and end of each day to ensure that all foreign substances are removed.
c) To train the employee to be aware of the hazards involved with foreign substances on the access system, and to stress that the employee should mount and dismount with extreme caution when the conditions cannot be corrected by the first two suggestions.

Enforceable standards

Currently there are no clear enforceable standards on requirements of means of ingress or egress from elevated vehicles. Guidelines can be obtained from the Society of Automotive Engineers' recommended practices SAE J 185 "Access Systems for Construction of Industrial Equipment." These guidelines are sometimes non-specific, vague or too lenient to provide adequate protection for operators or users of these vehicles.

For example, SAE J 185 allows thirty inches as a maximum height for the first step from the ground. It indicates that sixteen inches is the preferred height. Both dimensions are too lenient for safe step climbing. This SAE recommended practice is obviously providing the designer of the vehicle with flexibility needed to accommodate the various functions expected to be performed by such equipment. Such flexibility is often misused, misunderstood, or misinterpreted by designers as have been indicated by the review conducted for this study.

McCollum[12] (1980) even suggested that many manufacturers of equipment seem to ignore the SAE J 185 standards under the assumption they do not apply because these requirements are not directed to spe-

TABLE 1

Problem	Cause of Problem	Possible Solutions to Problems
1) Hidden steps not permitting visual contact by operator.	Lack of visual contact makes it difficult for operator to find next step which would lead to operator missing step or not gaining full balance on step.	To design an access, from the ground to the driver's compartment, with highly visible steps.
2) Lack of uniformity a) Step-to-step b) Vehicle-to-vehicle	a) Irregularity or non-uniformity in step design triggers perceptual error which causes operator to miss a step or lose balance. b) Lack of standardization of vehicular access/exit systems among vehicles used for similar tasks, or in same industry which may cause a perceptual error when operator uses an unfamiliar vehicle. This perceptual error may cause operator to miss a step or lose balance.	a) To design an access, from the ground to the driver's compartment, with uniform step size, shape, tread, and riser height. b) To design a standard access to be used on all vehicles that are used for similar tasks or for all vehicles used in the same industry.
3) Excessive height of step	Steps too high for user. User is unable to make whole-foot contact; user center of gravity falls outside of stabilizing forces causing loss of balance; user is subject to a pushing moment of force away from vehicle, and at times three-point method cannot be used.	First step height should not exceed twelve inches from ground. Each additional step should be at a uniform distance of twelve inches.
4) Lack of, or poorly placed side steps	Lack of side step or mislocation of side step forces the operator to reach beyond his or her physical limitation which causes a moment of force on the operator in the direction of the reach.	To install a platform or reduce distance of side step.
5) Lack of, or mislocation of handrails	a) Lack of handrails forces user to use other surfaces as handholds which may be inadequate. b) One handrail causes user to swing, forces user to find another surface for handhold, disrupts natural "climb pattern." c) Mislocation of handrail forces user to "search" for handhold, or to reach outside of physical capability. d) Unstable Handhold—user is forced to use a swinging door or steering wheel as a handhold which causes a shift in center of gravity and a loss of balance.	To design a continuous handrail that goes along each side of the access and across sidestep platform, if necessary.
6) Other miscellaneous design problems	a) No access system. b) Unstable or unsecure steps. c) Obstacles, such as a bulldozer shovel, in path of operator while he or she is mounting or dismounting the vehicle.	a) Every vehicle must have safe access b) Solidly fasten all steps and other components of the access/exit system c) Design an access/exit system that is obstacle-free.
7) Poor tread surface conditions	Poor tread causes loss of friction which increases chance of slippage.	Material should have a high coefficient of friction; require user to wear appropriate work shoes.
8) Environmental conditions	Foreign substance such as dirt, water, snow, ice and grease causes loss of friction and increases risk of slipping.	Design steps with drainage holes; inspect steps and clean them daily; train operators to mount and dismount with extreme caution under adverse conditions.
9) Lack of enforceable standards	Standards are non-specific, vague or too lenient to provide adequate protection.	Legislature standards directed to access/exit systems of all types of industrial equipment.
10) Lack of training	Operator may act in a careless manner during mounting or dismounting which may lead to an accident.	Train and supervise employees. Design to protect from human error.

cific pieces of equipment; an incorrect assumption. OSHA and ANSI have several standards that address ladder and step designs, access to cranes, requirements of elevated structures, etc.; all are designed to provide safe means of ingress and egress. Yet many manufacturers still seem to ignore them in their designs.

The solution to these problems would be to establish more specific guidelines for manufacturers or to legislate enforceable standards for the design of safe access/exit systems from elevated vehicles.

Lack of training

Sparrell (1980) suggests that the primary cause of accidents in his study was that operators are not aware of the three-point mode of descent contemplated by the authors of SAE J 185, or the high risk of injury arising out of failure to dismount in the correct manner.

In analyzing the potential causes of accidents, it is evident that the operator could not be considered the primary cause of the accident occurring, but rather the design of the access/exit system. As far as the operator being unaware of the three-point mode or the risk of injury, it is obvious that adequate instructions and training of operators on proper methods of mounting and dismounting of vehicles should prove useful in reducing the risks involved.

Human errors can be reduced by proper training and proper supervision. However, engineering design failure to recognize human limitations and needs and to provide adequate protection for safety should not be condoned. The state of the art in ergonomics/human factors provides us with ample information needed to avoid the types of equipment design errors illustrated.

Conclusions

Falls from elevated vehicles are serious and costly accidents. The poor design of the access/exit system is a major cause of accidents, because the design fails to recognize operator's human characteristics, physical limitations, expectations, and responses during a variety of situations.

Another problem is that the standards are inadequate, vague, or non-specific. Standards are sometimes overlooked because of ignorance of their existence, or incorrectly assuming that they do not apply.

Some possible solutions as suggested in Table 1 could easily be implemented provided that the designers and employers accord this problem the serious consideration it deserves. ✚

References

1. Miller, James M., "Efforts to Reduce Truck and Bus Operator Hazards," *Human Factors,* Vol. 18, No. 6, Dec., 1976.
2. McPeak, John S., "Summary Analysis of Powered-Industrial Truck Accidents," Department of Industry, Labor and Human Relations, Madison, WI, 1976.
3. Sparrell, C. F., "Off-Highway Vehicle Accidents and Operator Training," Society of Automotive Engineers, Inc., Warrendale, PA; Sept., 1980.
4. NIOSH, "A Human Factors Analysis of Materials Handling Equipment," Morgantown, WV, January, 1978.
5. Hall, Henry, "A Case Study—Falls From Transit Mix Vehicles," *Journal of the American Society of Safety Engineers,* June, 1967.
6. Szymusiak, Susan M. and Joseph P. Ryan, "Prevention of Slip and Fall Injuries" (Parts I, II), *Professional Safety,* June, July, 1982.
7. Society of Automotive Engineers, *SAE Handbook, Part 2,* "On-Highway Equipment, Off-Highway Equipment," SAE, Warrendale, PA; 1977.
8. OSHA, "General Industry Standards (CFR 1910)," 1981.
9. OSHA, "Construction Industry Standards (CFR 1926)," 1979.
10. ANSI, "Safety Requirements for Fixed Ladders," ANSI 14.3, 1974.
11. NASA, "Anthropometric Source Book Volume II: A Handbook of Anthropometric Data," *NASA Reference Publication 1024,* July, 1978.
12. MacCollum, David V., "Critical Hazard Analysis of Crane Design," *Professional Safety,* Jan., 1980.

Additional references

NIOSH, "Health and Safety Guide for Concrete Products Industry," Cincinnati, OH, June, 1975.

Woodson, Wesley E., *Human Factors Design Handbook,* McGraw-Hill, New York, 1981.

Confined space entry
Can the deaths and injuries be eliminated?

by Steve P. Krivan

Each year there are numerous deaths and injuries that occur during continued space entry activities within industry.

The following is a brief illustration of the seriousness of the problem related to confined space entry deaths and injuries as reported in various newspapers throughout the United States.

MARCH, 1981, CAPE CANAVERAL, FLORIDA—An aerospace worker died and four others were hospitalized following an accident which occurred when five workers, without breathing apparatus, entered the spaceship Columbia's engine compartment before it had been cleared of the pure nitrogen atmosphere used during a test.[1]

APRIL, 1981, AVONDALE, LOUISIANA—Two workers were found dead apparently asphyxiated by a gas that displaced the oxygen in the cargo tank they were inspecting at a shipyard.[2]

APRIL, 1981, NEWINGTON, N.H.—Two tank cleaners were presumed killed in a powerful explosion within an empty air force jet-fuel storage tank.[3]

JUNE, 1981, HARLAN, KENTUCKY—Three men were killed and at least two were injured in an accident at a coal company. A U.S. Mine Safety & Health Administrator Supervisor reported that the miners cut into a closed-off section of an old mine and were asphyxiated by black damp.[4]

JULY, 1981, NORTON, KANSAS—Three grain company workers died when they were overcome by molasses fumes within a storage pit at a grain elevator complex.[5]

DEER PARK, TEXAS (1980)—Two welders and a standby worker collapsed inside a vessel under repair and the welders both died before rescuers could aid them.[6]

FLOWER MOUND, TEXAS (1980)—Two 17-year-old workers of a municipal sewer servicing firm died while unsuccessfully attempting to rescue a 23 year old worker from asphyxiating on methane fumes in a city sewer lift station.[7]

BRADFORD, PA (1980)—Two workers died and two others were injured when a flash fire occurred in an oil tank car three days after a derailment, during the cleanup.[8]

One has only to read the newspapers to learn that confined space entry deaths and serious injuries are a somewhat frequent occurrence respecting no industry.

A report by Safety Sciences[9] prepared under contract for NIOSH tended to show that fatalities occurred more frequently in confined spaces. The National Institute For Occupational Safety & Health's criterion for a recommended standard—*Working In Confined Spaces*[10]—presents statistical data relating the number of "events," injuries, and fatalities from various data sources and shows the number of "events," injuries, and fatalities obtained for each of 15 basic accident and illness types. The injury/fatality statistics included within the NIOSH document reveals that the most hazardous conditions in a confined space are a result of atmospheric related events.

THE KEY ELEMENTS OF A CONFINED SPACE SAFETY PROGRAM

Each company, facility, utility and job site as applicable, should have an established confined space entry program that is well communicated with thorough training provided in the application of all phases of the program. The key steps of such a program include:

- A documented policy defining confined space.
- Hazard assessment procedures.
- Entry permit requirements and procedures.
- Adequate training.
- Program evaluation and enforcement.

1. Establish a confined space policy

A policy statement spells out the hazards of confined space entry and the procedures to be followed to assure safety. A permit system must be included within the policy to assure that all aspects of the procedure for safe entry are in place prior to entry. The permit system helps to assure that all concerned with the work are aware of the hazards involved and the precautions required. The policy and permit system development preferably solicits input and participation from the personnel involved with the implementation to assure that important details are not overlooked and to facilitate acceptance.

For our purposes, a *confined space* "is any piece of equipment, sump, trench, underground utility vault, sewer, etc. which must be entered through a manhole or restricted opening, having unfavorable natural ventilation which could contain or produce dangerous air contaminants, is an area in which exit during an emergency could be difficult, and is an area not intended for continuous occupancy."

A room could be considered a confined space whenever there is an activity involving toxic or flammable vapors, inadequate ventilation, and a source of ignition, all of which could present health or fire hazards to the room occupants. However, a confined space policy definition usually pertains to those areas where occupancy is not normally intended. Entry means placing the head or feet past the plane of a confined space entryway or other opening to the confined space. A confined space or area is also one in which dangerous air contamination cannot be prevented or removed by natural ventilation.[11] When a person works in this type of environment, the chance always exists that a reduced oxygen level, or combustible or toxic gases may be present.

2. Hazard assessment

No matter what type of confined space is to be entered, it must receive a hazard assessment. First determine what possible atmosphere contaminants could be present. Examine the process to see what materials are involved. Are raw materials, intermediates, by-products, or decomposition products involved which may contribute gases, vapors or fumes with toxic effects? Even if they are not lethal, are the contaminates irritating, do they cause dizziness, or in any way affect the ability to make sound decisions? Do they have adequate warning properties? In addition to assessing the hazards

---

RELATIONSHIP OF OXYGEN CONCENTRATIONS, DURATION OF EXPOSURE AND EFFECT

Concentration %	Duration	Effect*
20.9	Indefinite	Usual oxygen content of air.
19.5	not stated	Recommended minimum oxygen content for entry without air supplied respirators. Also OSHA standard 29CFR 1910.
18.0	not stated	ANSI Z117.1-1977 definition of oxygen deficient atmospheres.
16.0	not stated	Lowest limit of standards reported in literature.
12-16	seconds to min.	Increased pulse and respiration, impaired judgment, some coordination loss.
10-14	seconds to min.	Disturbed respiration, fatigue, faulty judgments, emotional upset, poor circulation.
6-10	seconds	Nausea, vomiting, inability to move freely, loss of consciousness followed by death.
Below 6	seconds	Convulsions, gasping respiration followed by cessation of breathing, cardiac arrest, death in minutes.

*Effect—Warning properties of low oxygen are inadequate except to trained individuals. Most persons fail to recognize danger until too weak to effect self-rescue. Signs include increased rate of respiration and circulation, which accelerates onset of more profound effects, including loss of consciousness, irregular heart action and muscular twitching. Unconsciousness and death can be sudden.

Figure 1

The hazard assessment should specify monitoring equipment to be used throughout the confined space entry procedure.

associated with the confined space, do not forget the hazards associated with the work to be done.

The confined space atmosphere must always be monitored to see if sufficient oxygen is present and to check for the presence of combustible vapors. Figure 1 summarizes the relationship of oxygen concentrations to duration of exposure and effect.

The hazard assessment should be used in deciding what additional monitoring is needed. In addition to the checks for oxygen content and the presence of combustible vapors, monitoring carbon monoxide, hydrogen sulfide, nitrogen dioxide, sulfur dioxide or other gases, vapors, fumes, or toxic agents, may be needed before the Confined Space can safely be entered.

An entry permit is the desired end product of the hazard assessment and the initial atmospheric monitoring and in reality it is valid only as long as conditions in the confined space do not change.

3. Entry permit/procedures

The entry permit is the heart of a confined space entry program. It brings together all parts of the program and documents the existing conditions in the confined space, conditions to be met before work in the confined space may begin and the requirements to be met while the work is being done. The permit also documents who is responsible for the various parts of the program.

Entry into a confined space should never be allowed without a permit issued by the proper supervision and with the appropriate safety and administrative back-up checks. A permit signed by the first line supervisor of the operating area, the person responsible for the hazard assessment monitoring, the supervisor of the personnel entering the confined space, and the individuals who are to enter the confined space provides the desired involvement. Since conditions in confined spaces can change, a permit is valid only for the duration of the shift, with the proper monitoring, or job, whichever is less.

The permit must be prominently displayed near the entrance to the confined space and specify:
- Duration of the permit.
- Identification of the confined space.
- Chemicals in vessel or confined space prior to entry.
- Procedure used to decontaminate confined space.
- Ventilation or other engineering requirements (i.e., blinding and lock-out required).
- Oxygen content checks and frequency if not continuous.
- Explosive gas checks and frequency of monitoring if not continuous.
- Short-term exposure limit and threshold limit value or company permissible exposure limits of potential contaminates from hazard assessment.
- Atmospheric concentration and frequency if not continuous.
- Radiation level checks and frequency of monitoring if applicable.
- Type of respiratory protection if required.
- Protective equipment required (gloves, boots, clothing, hearing protection, etc.).
- Safety harness/retrieval gear.
- Standby personnel.
- Tools specified for usage within the confined space.

Additional requirements may also be specified as a result of the hazard assessment for the confined space or the planned task. An entry permit is the desired end product of the hazard assessment and the initial atmospheric monitoring; in reality it is valid only as long as conditions in the confined space do not change. A study[9] of confined space injuries and fatalities conducted for the National Institute of Occupational Safety and Health,[10] indicates that a significant percentage of confined space related deaths and injuries were the result of atmospheric *conditions having changed after the initial monitoring.* This fact shows how critically important it is to *continuously monitor* for oxygen content, the presence of combustible gases, and the other possible air contaminants the hazard assessment suggest

Welding will generate fumes that can make a confined space work area unsafe in a short time.

may be present. As a minimum, periodic monitoring needs to be conducted throughout the confined space entry process.

It must be emphasized that the work to be performed within the confined space can cause conditions to change as well as present additional hazards (fall potential, electrical hazards, heat stress, etc.).

Ventilation—Continuous mechanical ventilation flow is most desirable throughout the entire confined space entry. Ventilation flow, calculated to meet the size of the confined space, can assist in meeting the comfort of the workers inside, as well as maximizing the safety of the interior air. Heat-stress can be avoided whenever planning includes ventilation for body cooling, sufficient work breaks, and appropriate body fluids replacement.

The work activity itself determines additional ventilation requirements as well. For example, welding will generate fumes that can make a confined space work area unsafe in a short time period. When freely moving exhaust hoods are used to provide control of fumes generated during welding, such hoods should maintain a velocity of 100 feet per/minute in the zone of the welding. The effective force of freely moving exhaust hoods is decreased by approximately 90 percent at a distance of one duct diameter from the plane of the exhaust opening.[10] Therefore, to obtain maximum effectiveness, the welder must reposi-

Although a confined space may have sufficient oxygen, and exhaust ventilation is provided for welding, respiratory protection for the welder may be in order.

tion the exhaust hood as he changes welding locations to keep the hood in proximity to the fume source.

A case history[12] illustrates the importance of the hazard assessment considering ventilation/respiratory protection requirements for the work being performed within a confined space.

Four employees of a local utility were repairing a water meter in an underground vault 18 feet x 6 feet x 5 feet with an opening 24 inches in diameter. To make the repairs, it was necessary to cut 26 cadmium plated bolts with an oxygen propane torch. Two men worked in the vault with one man cutting and the other standing beside him. Neither man wore a respirator and no ventilation was provided. Two other men remained on the surface. During the cutting of the bolts with the oxygen propane torch, a "heavy blue smoke" filled the vault. This smoke was exhausted after the cutting was completed.

The 56-year-old man who had cut the bolts died 17 days after exposure. He became nauseated shortly after the job and was seen by his family physician the next day for fever (102°/103°F), chest pain, cough, and sore throat. On the fourth day following the incident he was in greater distress and was hospitalized. Death occurred in two weeks and was attributed to massive coronary infarction and cor pulmonale.

The 29-year-old assistant complained of chills, nausea, cough and difficulty in breathing. He was treated for pneumonia and made a slow recovery. A reenactment of the work demonstrated that the exposure to cadmium was well above the threshold limit value.[12]

Symptoms attributed to cadmium poisoning include: severe labored breathing and wheezing, chest pain, persistent cough, weakness and malaise, and loss of appetite. The clinical course is similar in most cases. The injured frequently are well enough to work the day after exposure, but their conditions deteriorate until approximately the fifth day. At this point, the exposed worker will either get much worse or begin to improve.[13]

The lock-out procedures and blinding practices—the "zero-energy state" is another critical element within a confined space entry situation. Many companies have found multiple lock arrangements for electrical switches controlling energy to the confined space to be a good policy. Under this arrangement, there is a lock for each person within the confined space making it improbable for one person to activate equipment in the space.

Use good lock-out procedures.

Blinding practices must be thorough; control measures should include blanks that are inserted in the feed end of the separated steam/material, etc., flow lines into the confined space. A check list of all locations for the blinds assists placement and removal to be accomplished without overlooking any locations. Blanks should be of the same material composition as the feed lines into which they are inserted.

4. Establish training procedures

As a general rule the best policies and procedures are of little value unless they are communicated and employees properly trained to carry them out. Since the need for confined space entry may be limited, frequent follow-up or periodic *checks* may be beneficial.

Training materials and programs

Basic equipment required for safe confined space entry.

should be structured to encourage trainee involvement; simulations, dry-runs, and role playing are all useful techniques. In addition, they also allow a knowledgeable instructor to evaluate a trainee's skill and knowledge level without formal testing.

Topics to consider when planning a training program for work in confined spaces include:
- Task planning.
- Hazard recognition and assessment.
- Use and maintenance of monitoring equipment.
- Use of various types of respirators and other protective equipment.
- Responsibilities of the standby watch.
- Confined space rescue.
- Usage of appropriate engineering controls.

We will amplify on the responsibilities of the standby watch and confined space rescue.

Responsibilities of the standby watch—The manhole or standby watch person plays a vital role throughout the entry procedure. The watch person must be thoroughly instructed and supervised to be in constant view or communication with the individual(s) in the confined space. Watch personnel must understand that they are not to enter the confined space should a mishap occur; the watch person's responsibility is to quickly sound the alarm and to get pre-arranged res-

> ... policies and procedures are of little value unless ... employees (are) properly trained ...

cue personnel to the site if an emergency occurs. Rescue equipment must be readily available and designated rescue personnel thoroughly trained in its use.

Confined space rescue—Rescue equipment includes standby captive air supply and retrieval gear. The retrieval gear available must be commensurate with the potential problems presented by the hazard assessment. In the case of a narrow, vertical opening, for example, wristlets and a mechanical hoisting device would be needed to complement the retrieval harness being worn by the individual within the confined space and for usage with the hoisting line. The captive air supply (air supplied or self-contained breathing apparatus) must be suited to meet the opening and space to be entered. To illustrate, a narrow opening may not accommodate a self-contained breathing apparatus for a would-be rescuer and this should be discovered during the planning stages. In other situations, air lines may be impractical due to distances and obstructions within the confined space.

Unexpected entry—In addition to planned confined space entry, unexpected entry into confined spaces may occur and thus all employees within reasonable consideration should be trained to the hazards of confined space entry. An example of "unexpected" entry is where an employee may enter a confined space to retrieve an article inadvertently dropped into the confined space but fails to realize the hazards that may be associated with the confined space. A general educational effort can eliminate, or at least minimize, such an occurrence.

5. Program evaluation and enforcement

As in any successful safety activity, confined space entry procedures must be evaluated periodically to assure that all elements of the program are complete and are in fact working as intended. A minimum review is once annually, but preferably the program is reviewed more frequently because of the critical nature of confined space entry. Confined space entry "mistakes" tend to be unforgiving and generally result in one or more fatalities when the mistake is related to the atmospheric content of the confined space.

Strict enforcement of procedures is fundamental for continued safety success for confined space entry activities. The employees' awareness of the seriousness to follow the procedures consistently will, in a large measure, be based upon the seriousness with which he/she perceives management to attend to this area. If frequent checks are made and exact procedures are insisted upon, then all of the given training will have maximum benefit.

Training for the task

Can the deaths and injuries from confined space entry situations be eliminated? The answer is definitely yes. In order to accomplish this, however, the key elements within a confined space entry program as summarized above must be in place, well communicated, and *enforced* with appropriate training given periodically. Even with all of that firmly in place, a thorough analysis and follow-through of the hazard assessment of *each* confined space entry situation must be accomplished to assure that appropriate safeguards are in place for the specific confined space entry task.

Author's Note:
This paper has been written for the purpose of alerting safety personnel of the need for thorough and consistent confined space entry precautions, as well as to alert of the frequent loss of lives that have been occurring from confined space entry activities. The limitations for a paper of reasonable length on this subject precludes further elaboration upon the desired control measures.

While the key elements above have been found to be successful within the author's experience, there are no doubt additional considerations that other safety and health professionals have found useful and may wish to share via professional publication for the benefit of all.

References

1. Houston Post, March 20, 1981.
2. Dallas Times Herald, April 30, 1981.
3. Dallas Times Herald, April 30, 1981.
4. Louisville, Kentucky Courier-Journal, June 4, 1981.
5. Dallas Times Herald, July 12, 1981.
6. Dallas Times Herald, July 27, 1981.
7. Dallas Times Herald, 1980.
8. Fire Engineering, July, 1981.
9. "Search Of Fatality and Injury Records for cases related to Confined Spaces" (NIOSH P.O. Number 10947) San Diego, CA *Safety Sciences*. February, 1978.
10. "Criteria for A Recommended Standard—Working In Confined Spaces," U.S. Department of Health, Education, & Welfare, National Institute of Occupational Safety & Health, December, 1979.
11. A Primer On Confined Area Entry, Malvern, PA Biomarine Industries, Inc.
12. Zavon, M. R., and D. C. Meadows. "Vascular Sequalar To Cadmium Fumes."
13. Beton, D. C., C. S. Andrews, H. J. Davies, L. Howells, and G. F. Smith. "Acute Cadmium Fume Poisoning," *BR. J of Ind Med* 23: 293-301, 1966.

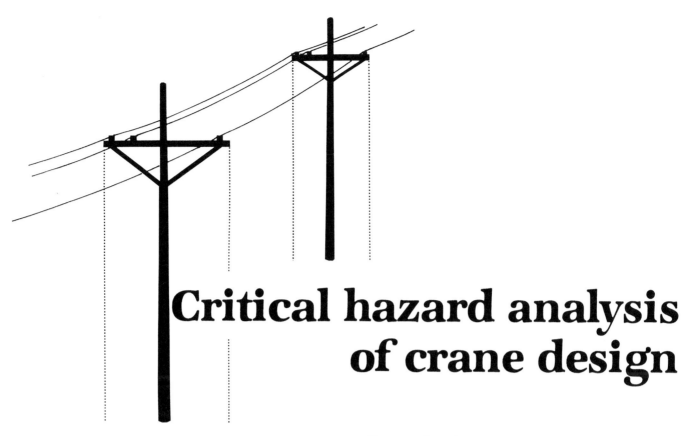

Critical hazard analysis of crane design

Far too often the cause of an accident is identified as "operator error," while "defective design" is completely overlooked. Defective design includes neglect of human factors considerations and foreseeable field use in a variety of environments.

A system safety approach will identify principal areas of defective design which seriously compromise the safety of mobile cranes. Design modifications will be discussed, with a historical summary of available technology for eliminating or minimizing each critical hazard.

Various failure modes will be analyzed with reference to general safety standards associated with each hazard, evaluating their applicability and effectiveness.

This paper is directed to the need for design improvement to overcome critical hazards which have not been effectively controlled by work-practice rules.

"The backbone of our construction industry... is the mobile crane."

by David V. MacCollum

For over a decade it has been recommended that system safety analysis, developed in the late fifties and early sixties to reliably assure for a quantitative safety factor, be applied more vigorously by industry at the time of design.[1,2,3]

The backbone of our construction industry and many of our industrial operations is the mobile crane. Little of what is available in system safety technology has been applied to crane design. Serious accidents involving these machines seem to increase as these machines grow in size and number.

An examination of a sampling of crane accidents quickly reveals that the cause of such accidents is generally identified as "operator error," while "defective design" is completely overlooked. Paramount in achieving greater safety in crane design is a shift of emphasis from reliance upon constant human performance to avoid accidents to an understanding of the reasons for variable human reaction. People accommodate hazards for seven basic reasons:

1. They are unaware of the hazard.
2. They are unable to evaluate the possibility of accident risk.
3. They work under competitive peer or production circumstances.
4. They work with machines which have incompatible controls, conflicting instructions, lack guarding, or lack warning.

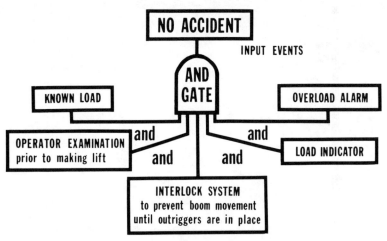

ALL SAFEGUARDS MUST BE REMOVED BEFORE AN UPSET CAN OCCUR

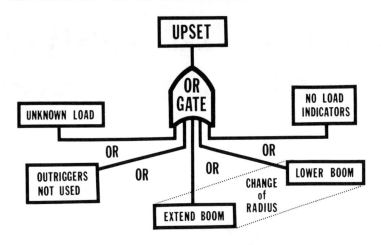

AS NO SAFEGUARD EXISTS TO PREVENT UPSET

5. They work with machines which exceed human performance.
6. They work with machines which create distractions (such as noise) or have visual obstructions.
7. They work under stressful environmental conditions.[4]

Progress in reducing crane accidents will only be made when priority is given to include safety in design to compensate and forgive for these foreseeable human errors.

Crane failure modes

The most frequently experienced crane failure modes in which the likelihood and severity of injury and damage has been found to be high can be broken into thirteen categories. These include:

1. Overloading, 2. Side pull, 3. Outrigger failure, 4. Hoist limitations, 5. Two-blocking, 6. Killer hooks, 7. Boom buckling, 8. Upset, 9. Unintentional turntable turning, 10. Oversteer/crabbing, 11. Control confusion. 12, Acess, 13. Unintentional powerline contact.

Problems associated with cranes can best be understood by examination of each of these thirteen categories of crane accidents to show the interaction of machine failure, man failure, and a mismatch of machine and man, all within the many variables of environment.

1. Overloading. It has been long recognized that the margin between tipping load and rated capacity for a given boom length and angle provides an exceedingly narrow margin of safety.

In 1954 this problem was addressed by the Portland District Corps of Engineers.[5] At that time, according to the *American Standard Safety Code for Cranes, Derricks and Hoists,* ASA B30.2-1943, paragraph 1311a, truck cranes required only a 17½% margin of stability above rated capacity for a tipping load. Thus, for a 30-ton rated capacity at a given boom length and angle, the safety factor was 5.25 tons, or a tipping capacity of 35.25 tons. This rated capacity was 85% of tipping load, or a 15% safety factor. This safety factor was found to be too narrow. After study, the Portland District Corps of Engineers raised the safety factor to 25%, requiring rated capacity to be 75% of tipping load. This was adopted by the Corps of Engineers and included in its *General Safety Requirements, 1958.*[6] No increase in the margin of safety required under USAS B30.5-1958, *Crawler, Locomotive and Truck Cranes,* and American National Standards Institute standards was made at this time. ANSI B30.5-1973, *Mobile Hydraulic Cranes,* still does not reflect any higher margin of safety. The Society of Automotive Engineers (SAE) in SAE J987, *Crane Structures, Method of Test,* and SAE J1063, *Cantilevered Boom Crane Structures, Method of Test,* addresses the method of conducting the test and the data to be recorded, but fails to define limits between tipping load and recommended safe operating capacity.

The operator of a crane should have the benefit of a 25% safety factor rather than the narrow 15% requirement. Operators should also have the benefits provided by load moment devices which are designed to prevent overloads from occurring. Several well-known manufacturers have developed and marketed such devices. These should be included on cranes as standard equipment.

Another method of reducing crane accidents from overloads is by installation of load measuring and load moment warning devices. Army, Air Force, and Navy Military Specifications commonly require these devices on mobile cranes. This approach has minimized crane upsets because of overloads in the crane operations of these organizations.

While moving a load as a traveling carrier, a crane can easily develop dynamics which can unexpectedly pitch the crane forward in an upset and can violently pitch the operator out of the cab. A "quick release" of the hoist line to increase the safety of the machine should be possible. The inclusion of this feature is recommended in some operating engineers' handbooks, but is absent from ANSI standards.

2. Side pull. Side pull or lateral loading can easily buckle a boom.[7] Many crane operations involve two cranes on a lift, and lateral loading can be unknowingly applied when the load must be turned. Usually operators are completely unaware of the hazard potential which is undetectably created within a "normal" lifting process. A detection system is needed to warn operators of side pull.

3. *Outrigger failure.* The majority of crane upsets have occurred when outriggers were not in place. Because use of outriggers is voluntary and left to the discretion of the operator who might not perceive a potential hazard, many very serious injuries and deaths have occurred. When use of outriggers is voluntary, an oversight on the part of the operator is foreseeable.

A mobile, rough-terrain crane is completely unstable on a side lift, and even a small load over to the side can overload a crane without outriggers in place. As the crane boom rotates on the turntable, the overload occurs so quickly that the operator cannot perceive the loss of stability until too late.

A few aerial basket trucks have hydraulic systems with interlocks which preclude boom operation until the outriggers are fully extended and fully supporting the crane with wheels completely off the ground. An interlock to prevent the boom from being swung to the side when the outriggers are not in place is a much needed design improvement.

Some cranes allow the rear outrigger to disengage from the pad when the load is first lifted. When the crane cab-boom swings around 180°, the outrigger can slip from such an unfixed pad connection and can easily cause the boom to buckle beyond the control of even the best operator. The installation of effective boom stops creates an additional margin of safety for controlling external forces to the boom and includes control over minor side lurch. Failure mode 8. *Upset* will expand on this topic in both travel and static modes.

4. *Hoist limitations.* On cranes, the hoist drum is outside the view of the operator. If the "headache" block is replaced with a lighter one, or if for any reason the line is not taut, the line can coil or hang up and not reel in safely. A guide or spooling device is needed, since the operator cannot see what is happening. Such a hang-up can cause the hoist line to be unintentionally parted.

"The operator ... should have the benefit of a 25% safety factor ..."

5. *Two-blocking.* Two-blocking is a hazard which occurs on both lattice-boom and telescoping-boom cranes. It occurs more frequently on hydraulic telescoping booms because the hoist line is operated by one control and the boom extension by another. Generally, the hoist line has a mechanical linkage to the power source which tends to provide a safety factor of lugging the engine prior to a parting of the line. This warning feedback to the operator gives an opportunity to avoid two-blocking. Two-blocking can easily occur while extending the boom because the hydraulic pump works quietly and surreptitiously and gives no warning of stress as does a lugging engine. An anti-two-blocking device as standard equipment is essential. ANSI standard B30.15-1973, *Mobile Hydraulic Cranes,* paragraph 15-1.3.2d, "Two Blocking Damage Preventive Feature," reads as follows:

"On a telescoping boom crane, with less than 60 feet of extended boom, a two-blocking damage preventive feature shall be provided capable of preventing damage to the hoist rope, and/or other machine components, when hoisting the load, extending the boom, or lowering a boom on a machine having a stationary winch mounted to the rear of the boom hinge."

Human factor's logic tells us that when an operator must use two controls, one for the hoist and one for the hydraulic boom extension, chance of error is increased. For many reasons an operator can forget to release (payout) the load line when extending the boom. When this occurs, the line can be "accidentally" broken. If the load line, when it breaks, happens to be supporting a worker on a bosun's chair, or several workers on a floating scaffold, a catastrophe can result.

Present standards basing use of a two-blocking damage preventive feature upon the length of the boom are ridiculous. From a human factor's standpoint, the device should also be standard for longer booms, since the longer the boom, the harder it is to see if it is in a two-blocking mode. I have been told by those opposed to use of such devices on larger cranes that because they cost more money, larger cranes have more competent operators. This position is not logical to me. A number of lawsuits have highlighted the inadequacy of these safety standards which fail to require an anti-two-blocking device.[8] MIL standards developed by the Navy Yards and Docks generally require an anti-two-blocking device, and the incident of accidents resulting from this cause is remarkably lower.

Anti-two-blocking devices should also be standard equipment on lattice-fixed booms. When walking a crawler crane with a long boom, a great deal of whip is created. The headache ball and empty chokers can drift up to the boom tip when walking a crane without a load. While the crane operator is busy watching the pathway of travel to avoid rough ground which violently jerks the crane, he is not watching the boom tip. When the hoist line two-blocks, it assumes the weight of the boom and relieves the pin-up guys of the load. Then, if the crane crawler breaks over a rock or bump, the flypole action of a long boom is sufficient to break the hoist line. (Lyles v. American Hoist, Oklahoma City, $425,000 settlement during trial.)

Anti-two-blocking devices are generally electrically initiated. However, a mechanical device has been developed by The Ederer Company of Seattle, Washington. Another method of anti-two-blocking is to have a hydraulic cut-out so the hoist line automatically pays out while the boom is being extended.

6. *Killer hooks.* The lifting hook is a critical link in lifting a load. One load hoist block found on many cranes fits flat to the boom. When the boom is in a lowered position, the hook can swing into any position and the straps rest safely, deep in the throat of the hook; but when the

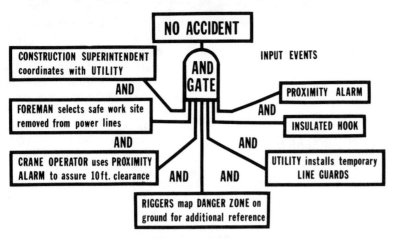

ALL SAFEGUARDS MUST BE REMOVED BEFORE AN ELECTROCUTION CAN OCCUR

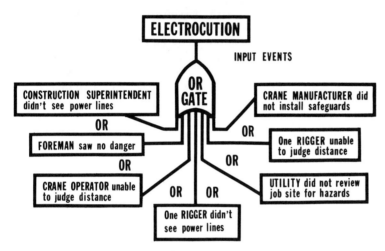

AS NO SAFEGUARD EXISTS TO PREVENT ELECTROCUTION

boom is raised in a nearly vertical position, the hook can rotate, allowing the straps to slide easily out of the hook. As a hook is subjected to wear, normal stresses gradually enlarge the throat, and this further increases the possibility of the straps' slipping from the hook. The common sheet metal safety latch becomes critical. There are a number of patented, positive type safety latches which prevent the hook from opening and require depression of the latch pin or lever before the hook opens. A leveling device on the block allows the block to remain always perpendicular, thus preventing a killer hook syndrome which can drop a load unexpectedly. The anti-two-blocking device mentioned in the discussion on two-blocking device mentioned in the discussion on two-blocking would also preclude holding the lifting block flush to the boom and activating a killer hook.

7. *Boom buckling.* In the 1950's boom stops were recognized as being essential, and many cranes are equipped with these devices as standard equipment. Simple, mechanical rests are not always the most efficient. In 1950 the Portland District Corps of Engineers also analyzed the concept of hydraulic snubs to absorb the shock and found "Boom Snub" to be a most efficient safeguard.

Use of boom stops avoids inadvertent pulling over (winching) of a nearly vertical boom with the winch line (two-blocking) or raising the boom beyond a safe angle. Boom stops also help avoid the forces created by a sudden release of the load, and also any forces created by wind.

OSHA does not require boom stops on cable-supported booms, and SAE J220 provides minimum performance standards, but not in terms of rated capacity as found in the Portland District Corps of Engineers' requirement of nearly thirty years ago.

8. *Upset.* Mobile, rough-terrain cranes can tip over very easily in the travel mode and crush the "tin" weather cab between the boom and the ground. This is a very serious hazard, and this type of crane should have rollover protection (ROPS).

Because of unstable footing or a faulty hook arrangement, a load may fall back on the crane and the operator is often in a very vulnerable position. A substantial cab is needed to protect the operator from such an occurrence.

As previously mentioned in Failure Mode 3. *Outrigger Failure*, the majority of upsets can be attributed to the operator's failure to extend the outriggers before commencing lifting operations. An analysis of some 300 crane accidents shows that half of these incidents occur when the crane operator either swings the cab or extends the telescoping boom. Both of these actions will rapidly increase the lifting radius and cause upset. This problem is a predictable human failure that deserves limit switches for lifting as mentioned in the earlier section. This critical, predictable human error happens at approximately the rate of once for every 10,000 days of crane usage which, in terms of reliability, would be unacceptable for mechanical failure, and, therefore, this human failure should be given equal priority for a safeguard.

The problem of simple imbalance also occurs when a long boom is in a vertical position and has a heavy counterweight. When the cab swings without outriggers in place, the counterweight is enough to tip over the crane.

9. *Unintentional turntable turning.* Pilots of planes with rectractable landing gears have more protection than do crane operators. If a pilot starts into a landing sequence and the landing gear is not down, an alarm gives warning that this critical function has been overlooked. The crane operator in a mobile, rough-terrain crane does not have this same type of safeguard to warn when the pin is not in place, locking the cab in position. A simple interlock to avoid this most terrifying sensation for a crane operator is needed.

10. *Oversteer/crabbing.* In some four-wheel, rought-terrain cranes, rear-wheel drive is actuated by a knob extending out from the dash, one-half inch from the right of the

steering wheel. This knob can be accidentally displaced by the right hand on the steering wheel while turning to the left, causing the machine to move crabwise sharply to the left. The knob can be brushed downward when turning to the right, causing the machine to turn sharply to the right with half the normal turning radius.

11. *Control confusion.* The arrangement of controls on the floorboard of cranes needs to be standardized.

There should be adquate space between the pedals to avoid overlapping from one pedal to the next when the operator wears large boots.

Hinge and lever pedals require different movements and can give an operator warning if the wrong pedal has been selected. This gives the operator a "sense of touch" in knowing that his foot is on the proper control. Think of the chaos which would result if foot controls varied from one make of car to another. Crane operators who must use several different types and makes of cranes must cope with such a situation when moving from job to job.

12. *Access.* Footholds used by operators to gain access in and out of equipment are often covered with oil from leaking hydraulics. This is an open invitation for serious falls.

Safe access is outlined in SAE J185, *Access Systems for Construction and Industrial Equipment,* first published in 1970. Many manufacturers of equipment ignore these standards under the assumption they do not apply because these requirements are not directed to specific pieces of equipment. This assumption is incorrect. Providing safe access for operators of equipment is necessary to eliminate another source of serious injury.

13. *Unintentional powerline contact.* A thorough review of existing statistics shows that some one hundred or more fatalities and several hundred cases of very serious/permanent injuries have resulted annually from contact with powerlines. This is by far the most serious of all types of crane accidents.

In the past a number of safe work practices have been implemented to prevent this type of injury. Reliance upon such work practices has been proven ineffective, as death and injury from this source continue.

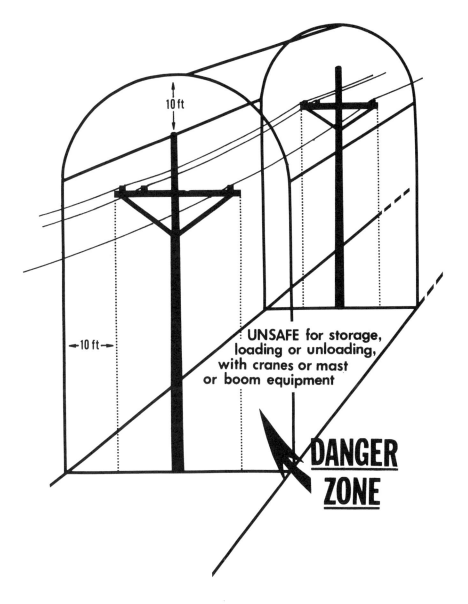

Robert Jenkins, now retired as Safety Director for the U.S. Army Corps of Engineers, investigated the reasons for this high incidence of severe injury in 1950. At that time he conducted a number of tests at Fort Belvoir to ascertain the ability of crane operators to estimate safe clearances from powerlines, visually. He found that dangerous misjudgments could be made in estimating clearance from a small wire against a bright background. Unfortunately, these important findings were not recorded. However, Mr. Jenkins presented testimony on these findings in 1973 in the Cook County Circuit Court in the case of Burke v. Illinois Power, which was the basis for a decision that crane manufacturers may be strictly liable for failing to equip cranes with known safety devices, such as insulated hooks, boom cages, and proximity alarms. This verdict of $2,500,000 was upheld on January 18, 1978, on appeal.

Since that time a number of cases have been heard concerning the need for such safety devices.

In order to examine this area of human factors as related to the ability to judge clearance distances from powerlines, the author and Dr. Lorna Middendorf conducted a field study as a laboratory exercise for the students at a Green Shield Professional Development Conference sponsored by the National Farmers Union Casualty Companies. The results of this field study were published in two papers presented at the 22nd Annual Meeting of the Human Factors Society.[9,10]

A definite trend is making itself apparent that a system approach to safety is essential in which safeguards can be implemented by each area of involvement. Courts are recognizing that everyone has a res-

> ... Costly governmental programs could benefit greatly by providing safer work environments which forgive known variables in human behavior.

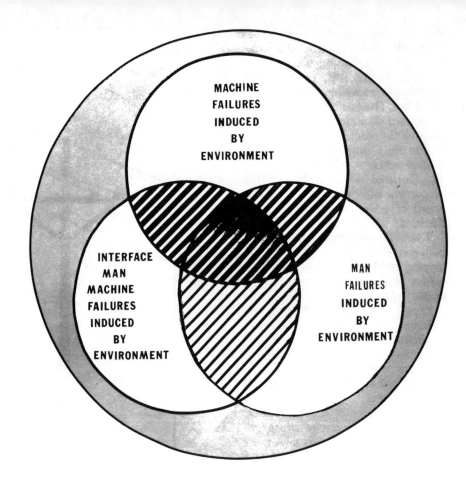

ponsibility in these types of very serious accidents—crane operators, riggers, utilities, contractors, and crane manufacturers. It appears that the courts are holding crane manufacturers to a greater degree of safety care than in the past.

Analysis of accident litigation provides a positive insight into the complexities of applied system safety approaches. System safety is often used as an analysis method for the benefit of the court. When referencing various safety standards it has been found that they are incomplete or state only that certain hazards must be considered. For instance, the National Electrical Safety Code requirements for safe clearance around bare electrical conductors should warn designers of cranes that these lines also create a hazard for cranes.

Conclusion

This critical design analysis of cranes was developed to show how system safety analysis can be a viable design review in industrial applications. It is unfortunate that use of the system safety approach has not been adopted to any great extent to provide a higher degree of safety in exceedingly high risk operations. Hopefully, this expertise will be applied to industrial applications during the next decade to provide the insight which will substantially reduce the high incidence of severe injuries and deaths which is occurring with increased use of production equipment.

I believe that costly governmental programs such as OSHA could benefit greatly by redirecting emphasis to creating incentives for providing safer work environments through **design innovations which forgive known variables in human behaviour.** ✚

From a presentation made at the System Safety Society's 4th International Conference in San Francisco, July 9-13, 1979.

References

1. MacCollum, "Reliability—A Quantitative Safety Factor," *Professional Safety*, May, 1979.
2. MacCollum, "A Systems Approach for Design Safety," *Professional Engineer,* November, 1968.
3. MacCollum, "Testing for Safety," *National Safety News,* February, 1969.
4. MacCollum, "Freak Accidents," *Hazard Prevention,* Sept/Oct., 1978.
5. MacCollum, "How Crane Load Tests Prevent Accidents," *Pacific Builder &* Engineer, March, 1957.
6. *General Safety Requirements,* EM 385-1-1, March 13, 1958, para. 18-4.
7. Brolin, C. A., "Destructive Testing of Crane Booms," *National Safety Council Proceedings,* October, 1977.
8. MIL-C-28622 (YD), Crane-Hydraulic, Wheel Mounted 4 by 4, Full Revolving, 45 Ton, para. 3.6.2, Boom, Sept. 30, 1971. Also see MIL-C-28614A, June 12, 1975, and MIL-C-281175C, March 28, 1977.
9. MacCollum, "How Safe the Lift?" 22nd Annual Meeting Proceedings of Human Factors Society, Detroit, MI, October, 1978.
10. Middendorf, Dr. Lorna, "Judging Clearance Distances Near Overhead Powerlines," 22nd Annual Meeting Proceedings of Human Factors Society, Detroit, Michigan, October, 1978.

Machine safeguarding and safety management

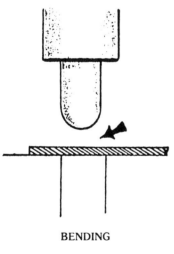

Figure 1. Point of operation hazards.

by D. A. Colling

The importance of machine guarding in industrial safety programs is well recognized. OSHA regulations, CFR 1910, Subpart O, specify that: "each employer shall install, safeguard, operate and maintain at all times... machinery in a manner which protects all employees against the hazards of moving parts, flying material, falling material and inherently hazardous material."

Nevertheless, workmen's compensation cases and third-party products liability cases are preponderant with traumatic injuries and amputations suffered by employees in the workplace because of unguarded or improperly guarded machinery. The ramifications of these injuries have already influenced products liability and workmen's compensation laws,[1] particularly in the areas of contribution and indemnification.

Contribution and indemnity became issues in the safeguarding case of Skinner v. Reed-Prentice when the manufacturer filed a third-party complaint against the employer seeking contribution because of negligent acts and omissions of the employer with regard to safeguarding of the subject machinery. Contribution, in the legal sense, simply allows the manufacturer to be reimbursed by the negligent employer for the monetary award to the injured employee. When permitted, contribution costs are borne by the employer, not by the workmen's compensation carrier.

Although an employer may not be subject to third-party tort claims in all instances because state laws vary, the potential of a claim in itself is impetus for improved safety programs. Accidents are rare events and, as such, do not necessarily aid an individual corporate safety program.

Examination by the author of the circumstances of a number of accidents involving machine safeguarding has permitted some insights which may provide guidance to corporate safety professionals. The machinery includes punch presses, press brakes, printing presses, shears, conveyors, and other specialty machines.

The hazards of machinery

Of the four hazards enumerated by OSHA regulations—moving parts, flying material, falling material and inherently hazardous material—it is the author's experience that most machinery and machine safeguarding accidents are associated with moving parts.

The hazards of moving parts in machinery are two-fold: those at the point of operation, i.e., where the intended work of the machinery is actually performed during processes such as punching, forming, or cutting, and those at pinch points other than the points of operation at which a part of the body may be caught. Examples of these types of hazards are shown in Figures 1 and 2. Of the industrial accident investigations examined for this report, about 40% involve pinch points and the balance point of operation.

Hazards in moving machinery cannot always be simply visualized or categorized as in Figures 1 and 2. In a recent article,[2] it was pointed out that many hazardous conditions are not properly controlled because they are not recognized as hazards. This author has found inability to recognize hazards in moving machinery accidents to be a contributing factor to accident and injury in about 30% of the cases examined.

For example, in an operation where pressure rolls were used to apply liquid glue, much in the manner that paint rollers apply paint, a barrier safeguard was provided on the in-running nip point during application procedures. Normal maintenance cleaning procedures using solvent and wiping action, however, involved reversing the rolls, thus reversing the position of the in-running nip point which was unguarded. Accident and injury occurred during this normal maintenance operation.

Other difficult to recognize hazard problems in machinery guarding involve occasional use of the machinery, material transfer onto or from the machinery, location of controls with no visibility of activity at the pinch point or point of operation, operation of machinery for routine or non-routine maintenance, and wear in machinery parts.

Machine safeguarding methods

Principles of machine safeguarding not only consider protection of the operator at the point of operation and at pinch points, but must also consider maintenance activities and servicing or adjustment which has to be done dynamically. The methods are not new and have been described in OSHA regulations, ANSI specifications, professional society publications, and safety manuals; they are summarized in Figure 3.

The most common method of safeguarding that is practiced wherever possible, for reasons of effectiveness and cost, is the permanent barrier-type safeguard which enclosed the hazard. In most cases, manufacturers provide permanent barrier safeguards for point of operation and for power transmission components. However, in many instances where older equipment has been modified for use in new applications, well-meaning employers have made and installed various types of barrier safeguards. In these cases, it is most important that a safeguard be adequate.

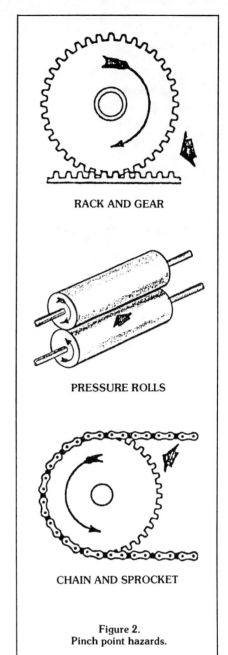

Figure 2.
Pinch point hazards.

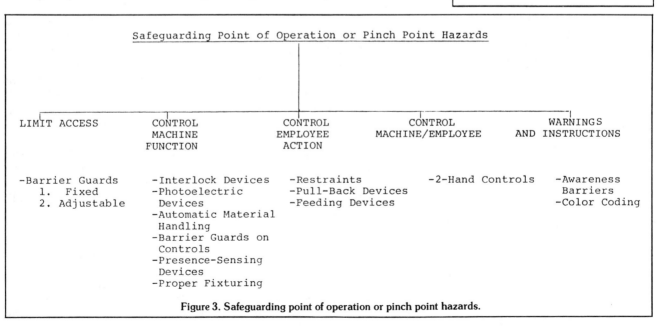

Figure 3. Safeguarding point of operation or pinch point hazards.

Figure 4 illustrates the relationships between guard location and opening which will prevent entry of the hands or fingers into the hazard area. In at least one of the accidents examined by the author, failure of the safeguard to conform to these design requirements may have contributed to the accident and injury.

Improving safety records for machine safeguarding

Machine safeguarding principles will probably not change drastically in the future, but the need exists for improving safety when working around machinery. Improving safety depends to a large extent on implementation of a systems approach to hazard control,[3] whereby hazard control is systematically integrated into the management process.

Management is a "process of accomplishing certain desired results or objectives through the intelligent utilization of human effort and physical resources."[4] With regard to machine safeguarding, principles of safeguarding as a hazard control are derived according to fundamental rules, but the process of how these principles are applied within an organization is not; this process is management.

The importance of proper management in machine guarding safety is exemplified by one incident in which an employee's hand became caught in the unguarded nip point of pressure rolls while making a dynamic adjustment. The rolls were located about six feet off the floor and the employee was standing on an unstable stool which tipped, causing his hand to enter the in-running nip point. Proper systems management would provide a stable stool or ladder for such adjustments before performing extensive modification of the machinery to properly safeguard it.

In examining the industrial accident investigations for this report, modifications *other than placement of safeguards* which would be made by a systems management process were considered. It is the author's opinion that a majority of the accidents and resulting injuries would not have occurred if the following corrective actions were made:

1. Better Hazard Recognition
   a) Eliminate multiple control points.
   b) Eliminate flexible controls for point of operation hazards.
   c) Improve lighting.
   d) Examine adequacy of emergency stops and correct, if necessary.
   e) Recognize worker stressors.
   f) Understand safety in rarely used equipment.
2. Better Training and Worker/Job Compatibility
   a) Institute formal training.
   b) Re-examine training needs.
   c) Periodically retrain employees.
   d) Select employees with safety in mind.
   e) Determine if there are "differences" in similar machines.
3. Better Inspection and Maintenance
   a) Examine for wear, which causes multiple tripping action.
   b) Inspect interchangeable parts which might fail in use because of wear.
   c) Inspect whether dies are secure.
   d) Maintain floors and keep

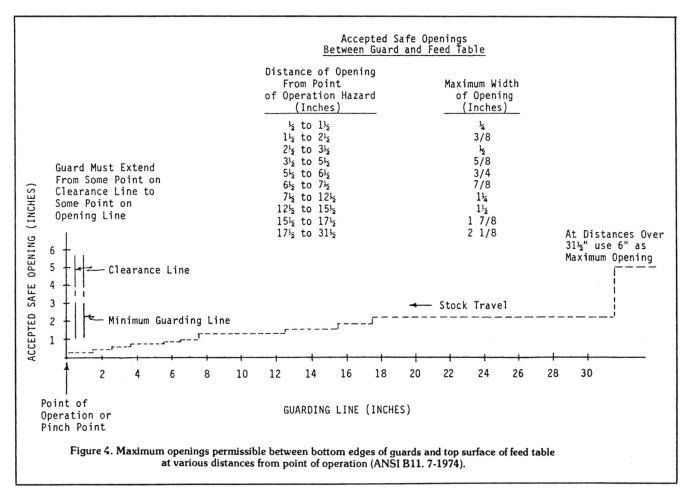

Figure 4. Maximum openings permissible between bottom edges of guards and top surface of feed table at various distances from point of operation (ANSI B11. 7-1974).

area clean and dry.
4. Improved Management Attitudes Where Improper Focus on Safety
   a) Legislate safety management.
   b) Mandate safety management.
   c) Educate safety management.

These four areas for corrective actions are not all new. For example, better hazard recognition has been aptly described[5,6] as has better training and worker/job compatibility which is in the category of human factors engineering.[7]

Two incidents which might have been prevented by these corrective actions are noteworthy.

The first is the rarely used machine, usually old, which can produce enough parts in one day for a month's or even six months' consumption in the plant. Such an operation invites disaster because any training is not retained, even by the most trusted employees. It is on such occasions that safety and hazard recognition must be the highest priority, perhaps assigning an engineer as well as the operator to the equipment while in use.

The second incident involved similar machines which were actually "different." The incident occurred on one automatic stapling machine of several in the department. On the subject machine, the trip stapling was positioned forward of all the other machines. Predictably, an operator reassigned to this machine, unaware of the difference, punctured a finger with a staple. Awareness of the difference and a simple warning by the line supervisor might have prevented the accident and injury.

Better inspection and maintenance procedures cannot be neglected. In many cases, the safety professional should involve the engineering department in this activity. For example, one of the incidents investigated was caused by failure of the safety device because of a worn, interchangeable part which fractured. This small part, if it had been examined closely before its use in the device, would have been discarded because of the obvious wear.

It is the last category of corrective actions which is the most perplexing. Good safety programs are impossible without the full endorsement of top management. Management

Better inspection and maintenance procedures cannot be neglected.

must recognize their role and they most probably do so in most large companies. The problem in management's improper focus on safety is more unique to small companies. In such cases, legislation such as OSHA has been ineffective. Mandates for safety through workmen's compensation carriers may be more appropriate. Insurers working in cooperation with safety educators may find a solution to this problem; this approach should be evaluated and implemented.

If nothing is done, there are indicators that forecast some expensive lessons in safety attitudes. A recent decision in the California Supreme Court involving asbestos[8] circumvented the workers' compensation law because the employer fraudulently concealed the hazard from the employee and induced him to continue to work under hazardous conditions. The time will come soon when some ingenious plaintiff's counsel will apply such reasoning to an accident involving unguarded or improperly guarded machinery!

References
1. Jeffers, S. A., "Skinner v. Reed-Prentice: Its Effect on the Doctrine of Contribution and Indemnity as Applied in Illinois Workmen's Compensation Third-Party Actions," *Southern Illinois University Law Journal,* pp. 556-578, 1978.
2. Boylston, R. P., "Recognizing Hazards: Whose Responsibility?" *Occupational Health and Safety,* Vol. 50, No. 9, pp. 94 ff, Sept. 1981.
3. Capps, J. H., "Systems Concepts for Safety Progress," *Professional Safety,* Vol. 26, pp. 41-45, March 1980.
4. Firenze, R. J., *The Process of Hazard Control,* Kendall/Hunt, p. 44, 1978.
5. Meagher, S. W., "Machine Design Mechanisms of Injury," *Professional Safety,* Vol. 25, pp. 19-21, Feb. 1979.
6. S. W. Meagher, "Designing Accident-Proof Operator Controls," *Machine Design,* pp. 80-83, June 23, 1977.
7. A. Chapanis, "Human Factors Engineering for Safety," *Professional Safety,* Vol. 26, pp. 16-21, July, 1980.
8. "Johns-Manville Products Corp. v. Contra Costa Superior Court," 612 *Pacific Reporter* 2d948.

Professional Safety:
Exposure reduction in LP-gas installations

Fire as result of leaving transport connected to storage after transfer operations.

by Hugh F. Keepers

In the early 30's and 40's liquefied petroleum gas, commonly referred to as butane, was just being introduced into the rural areas. Butane replaced coal and coal oil as a more efficient, clean burning fuel used for food preparation and home heating. Butane was referred to as "the gas system beyond our natural gas mains."

There was very little known about the characteristics of butane in the early days, except that it would burn and could explode if not handled properly. So by trial and error the LP-gas industry got its start. (Many times by error—chalked up as learning the hard way.)

As people in industry became more and more familiar with the product, more and more uses were discovered. When the petrochemical industry, with its vast knowledge and research capabilities, focused on this product they saw its potential and thus "butane" developed into one of the most versatile fuels known to man.

Today in Houston alone, there are over 6,500 different uses for butane and propane (LP) gas—from domestic use to carboration, steel manufacturing, glass processing, aerosol propellant, propane air-gas mix systems, large industrial plants, and standby systems to support peak operations at hospitals. Also for standby to large manufacturing plants where it is used to supplement depleted natural gas supplies during winter periods.

The LP-gas industry did not always have the wide acceptance it now enjoys because of the fear that earlier users had of the product. However, the industry realized that to sell the product to industry, they would have to remove fear and replace it with respect. This was done by selling safety along with the product.

City and state governing bodies across the nation adapted national safety standards for the safe construction of LP-gas vessels and wrote rules and regulations setting forth minimum requirements.

What are your exposures?

These rules and regulations were designed for a safe installation, under normal operations, i.e., that LP-gas would be safely transported and dispensed without danger to personnel, property, an adjoining neighbor, or the neighbor's property.

Today however, it is not enough to be content with these earlier rules and regulations. They lead to a false sense of security. A plant engineer, safety engineer, or safety director must be looking at exposure potentials now and 10 years ahead. He must exceed the minimum standards.

For instance a plant built with LP-gas storage 5 years ago in the middle of nowhere, now finds itself surrounded by sub-divisions, shopping malls, churches, schools and hospitals—known as exposures or loss potentials.

A fire in an industrial plant with LP-gas bulk storage could result in a container rupturing, especially if the vapor space of the container is exposed to prolonged heat and flame. I have personally measured the distance that some of these vessels have travelled following such a rupture and found that they can travel from ¼ to ½ mile.

You must also consider the concussion factor which always follows such incidents. These concussion forces can and do destroy buildings in the immediate and surrounding areas. The shock wave produces forces which move in an outward direction, building up as they move outward from the heated expanded gases expelled from the rupture.

What can we do to reduce the exposure?

Prior to selecting a new plant site, the plant engineer, safety engineer, or safety director, along with the architect and management, should meet with state and local officials to consider exposures, future development in the area, and future expansion of plant facilities. Often the architect custom designs the plant to meet the demands of management for production capabilities without consideration of exposure liability. Safety must be placed in our designs for the installation of an LP-gas standby unit, and the design should allow for additional storage that may be added in the future.

The following five items should be placed high on the priority list in making an LP-gas installation at an industrial plant:

Five top priorities

1. Never install LP-gas containers end to end. In a fire, if one container should fail, its trajectory could be directed into the end of the second container. All LP-gas containers should be placed parallel to each other, and should head in the same direction.

2. In designing fuel storage locations, make sure that the ends of the vessels are not directed toward buildings. A ruptured vessel usually travels in the direction the "head" is pointed.

3. Loading and transfer operations should be located at least 50 feet from the storage containers themselves and at least 50 feet from any open fired vaporizer. Should there be a leak of a valve or a transfer hose rupture, this distance could mean the difference in having time to shut off the valve before the gas reaches the open flame. These 50 foot distance requirements from the container help to limit the exposure when a transport or bobtail unit is on fire by being separated from the storage itself.

4. Concrete or steel bulkheads at transfer locations must be directed away from all storage facilities. Why? Should the driver drive off with the hose connected, the escaping gas if ignited, would be directed away from the storage containers.

5. All thermal acting automatic shut-off valves located at the concrete or steel bulkheads should be extended to a level above the top of the bulkhead, high enough to expose them to the quickest heat in case of a fire.

Too many times a bulkhead is designed with the thermal acting shut-off valve located behind the bulkhead in a position where heat from the fire cannot actuate the fuse which controls the shutdown valve. Considerable money is wasted if thermal valves do not work as designed.

Travel of ruptured vessel toward building.

Containers perpendicular or parallel to each other.

The LP-gas standby storage installation must exceed the minimum standards of today to survive the possible accidents of tomorrow and not add to the industrial plant's liability.

Proper training of maintenance and operating personnel in the characteristics of LP-gas and the necessary emergency steps to be taken to shut down the operation should be the safety engineer's priority. This training should also include local, municipal, and voluntary fire personnel who could be called to assist in an emergency. All personnel other than those assigned to such emergency shut down operation crews must be evacuated.

This shows overall catastrophe of an LP-gas industrial fire.

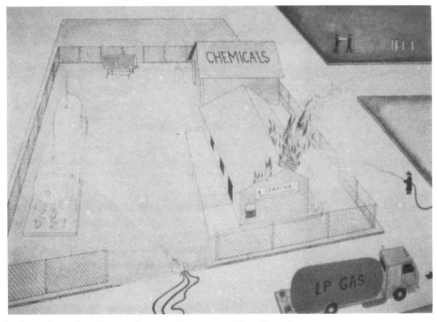

Cooling of vessel with water fog to reduce radiated heat from burning exposure.

Evacuation routes must be directed away from areas where emergency personnel and equipment will be responding. This removes congestion from the area, and saves precious time needed by emergency crews. Designated emergency routes to the plants must be established when preplanning to insure the quickest avenue to the storage area for emergency vehicles.

Don't put all of your eggs in one basket

This is a phrase we have all heard many times: Many industrial plants place priorities on production when planning a new plant site, leaving one of the most important parts of the operation to the last—fuel storage.

Knowing the hazard potential of fuel in general, the planners place it in one corner—any corner that is left after the buildings are erected. Then they try to cut corners on the state and local safety standards by saying, "Well, we don't have any other place to put it;" or "This is all the property we have." After placing the LP-gas standby system in one corner of the plant—which barely meets the minimum distance requirements—they find themselves backed by the joining property which could be developed at any time. They also continue to add storage of gasoline, diesel, lubricating oils, and even chemicals to this already hazardous area.

In the minds of some plant managers, anything considered hazardous should be placed in the area designated for dangerous commodities. They give no thought to their compatibility. No consideration is given to the fact that should there be a fire accident in this area, an explosion could occur, thereby setting off a chemical reaction of the other products stored there resulting in total disaster.

Fuels as well as chemicals must be stored with utmost care. This is an area where the plant safety engineer must be part of the internal and initial planning. He can then continue to monitor future growth of the plant, making sure that hazard control is a part of the corporate structure.

The safety engineer must also make sure that LP-gas storage is located away from other static fuels, such as above-ground diesel, oil, and gasoline tanks. Secondly, to insure that any leak in the static fuel tank would not flow under the LP-gas storage installation. Dikes, diversion curbs, or grading of the area are just a few of the ways to prevent such fuel from flowing toward or under an LP-gas installation. A leaking gasoline or diesel oil line can create a very hazardous condition and without necessary precautionary steps, the chain of events can result in total disaster to the plant.

When an industrial plant has installed a battery of large LP-gas storage containers at one location, and later decides to add containers, consideration should be given to locate the new battery of containers in a separate and remote area from the first storage location. Should one battery of tanks be involved in an accident, the plant could avoid a long shutdown by utilizing the secondary battery until necessary repairs can be made. More importantly this would reduce your exposure potential.

In an emergency

Emergency personnel can usually handle a small battery of LP-gas containers with success; however, when there are more than four containers in any given area, fire and emergency personnel cannot be effective as they might be on a small storage installation. This is due to the limited availability of water when

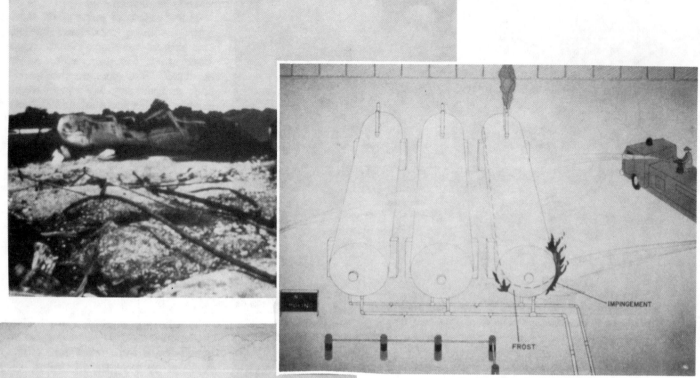

Cooling effect by emergency personnel on small batteries of LP-gas containers.

Separation of battery of containers.

a concentration is needed to cover all the vessels involved at one time.

When LP-gas tranport units or bobtail trucks are not being used, they should not be left parked side by side or next to a storage container. LP-gas trucks not in use should park at least 25 feet from each other. This provides protective space if one is involved in a fire, without destroying the whole fleet.

All liquid valves should be in a closed position on all LP-gas trucks unless the unit is transferring products. Leaving an LP-gas transport or bobtail truck connected to an LP-gas storage following the transfer of fuel adds to your exposure. A transport left connected means precious time could be wasted in disconnecting from the filler valve to move this unit to a safe location. Not realizing that the unit was still connected to the bulk storage someone might try to drive the unit off in an emergency situation and break the hose or fitting creating another fire and adding to the already bad situation.

Personnel without adequate knowledge of the seriousness of the hazard, permit spare LP-gas cylinders to be stored inside an already congested area thus creating another serious exposure in shops and warehouse complexes. A small warehouse or shop fire could result in total loss if the fire reaches LP-gas cylinders. Only cylinders properly mounted on an industrial forklift unit would be allowed inside the shop or warehouse area and only while the driver is in attendance. When a unit is not in use, it should be parked on the outside dock or ramp in a well ventilated area. The storage of these spare cylinders could prevent evacuation or exit from an area should a fire occur, and could likewise keep emergency fire service personnel from entering the building.

Department of Transportation (DOT) LP-gas cylinders have spring-loaded pressure relief valves in the vapor area which discharge in the event of increased pressure caused by temperature increase of the contents. All unnecessary exposure should be avoided. Areas outside of warehouse and shop buildings must be provided for this storage of spare cylinders. You should refer to the NFPA and other state and local standards for proper safe distance and location of such containers.

Lessons from 25 years of ROPS

by David V. MacCollum

In the last several decades, we have seen a drastic change in public priorities for safety. We, as safety professionals, need to examine the factors which support these priorities.

The rapid expansion of technology has created tremendous productivity and abundant benefits. Machines have become our modern-day "beasts of burden." But these benefits have often been diminished by the slowness exhibited by those involved in such technological advances either to apply safety technology at the time of design or correct inherent hazards as soon as such hazards become apparent.

Tractors are good examples of this syndrome. Had available safety technology been applied, there would have been a great deal less pain and suffering associated with these machines. Society should not be burdened in 1984 with death and crippling injury caused by hazards perceived in the early 1900's, especially when technology makes it possible to eliminate such hazards effectively. No one should be required to give up his or her life because a manufacturer decides it is too costly to eliminate a hazard that accident statistics have clearly shown to be the cause of frequent death.

Our technology can produce wealth for all, without subjecting a portion of our population to unreasonable risk. I feel collective industrial indifference delayed incorporation of what is now recognized as essential operator protection from rollover, falling objects, and other hazards. The parochial approach to safety to rely upon operator skills backfired into a confrontation of litigation, legislation, and public and community cost, plus lasting grief for many families.

We can perceive hazards, and we have the ability to overcome them at time of design and manufacture. The objective of this article is to present the case for safety engineering as a protector of industry, community and individuals.

Progress

The twenty five year history of the evolution of rollover protective systems (ROPS) for tractors gives insight into better ways to control the hazards that continue to kill and maim. It proves that more emphasis needs to be placed on safety engineering at time of design and on design improvement whenever equipment develops a history of repetitious accidents. The fact that death and injury from accidents involving tractor rollover have been substantially reduced because of the installation of ROPS make a case for greater design safety emphasis.

Definite progress has been made in the recognition of the need for inclusion of physical safeguards in design of equipment so that we do not have to rely solely upon human performance to avoid or minimize injury- or damage-producing circumstances. The slow acceptance of this safety philosophy has contributed to higher casualty rates than should have been experienced. The magnitude of human suffering, the horrendous cost, and the chaos that results when safety features are neglected at time of design must stop. Inherent hazards in machinery can be overcome in design. Safety engineering can insure the profitability of a product during its entire life cycle.

No product should have a hazard which causes death and injury for so long a period as the history of rollover has shown. From early recognition of the vulnerability of tractors to upset, through user development of sturdy brush guards that could withstand the forces of rollover, to ROPS, too many lives were lost that could have been saved had the available information been used earlier and necessary safeguards provided as standard equipment.

Need for protection

The need for some type of rollover protection was recognized over sixty years ago, even before the introduction of tractors into the marketplace. In the horse and buggy days, people riding in carriages, wagons, and trains were often killed or injured by falling objects, displacement by intruding objects, or rollover. These hazards were overcome by providing more substantial cabs and by sloping curved roads and railroad tracks with "super" to reduce the hazard of rollover on curves.

Since World War One, use of both wheeled and crawler tractors as general utility machines has become widespread. As the tractor population grew, it was soon found that tractor operators were being killed and injured by falling objects, by swinging loads, branches, or anything else which might intrude into the operator's compartment or station; by tractor rollover, or by being pitched off by erratic movements.

Although rollover did not occur as frequently as the other hazards, the resulting consequences were far more severe, usually fatal. Tractors were found to roll over backwards when pulling excessive loads or climbing steep slopes (particularly wheeled tractors), when backing over steep slopes, when attempting to ascend excessive grades. Most rollovers were lateral and to the side, but some were frontward, end-over-end when operating on or descending excessive grades. To guard tractor operators against these hazards, a few perceptive owners and users began to design, fabricate, and install protective canopies on tractors to cope with the dangers that they encountered.

Early in the history of tractor use, Edward Hewitt, an agricultural engineer, made a study of the principles of wheeled tractors and identified the hazard of rollover. His findings were published in the Society of Automotive Engineers (SAE) Proceedings in 1919.[1]

The forces required to cause rollover were calculated by Dr. McKibben, an engineer, in 1928, and he published several articles on the subject in the American Society of Agricultural Engineers (ASAE) Journal.[2] Since that time a number of accident summaries have been published by various states and by national organizations concerning the rising tractor death toll from these hazards.

Development of rollover protection

By the 1940's sufficient accident data had been gathered that indicated protective canopy guards were needed to protect operators from falling and intruding objects. In April, 1945, California led the way. In 1946 the U.S. Army Corps of Engineers began requiring brush guards on tractors. Oregon and Washington adopted similar requirements in 1948. Reference 3 contains the lists of the Corps of Engineers' *General Safety Requirements* and the *Logging Codes* of Oregon, Washington, and California.

No product should have a hazard which causes death and injury for so long a period . . .

In 1956 the University of California's Agricultural Extension Service at Davis developed a driver safety frame for wheeled farm tractors. During its studies, it was found that since insufficient time existed for the operator to dismount safely in the event of rollover or upset, the only safe alternate for operator protection was the driver safety frame.[4] About the same time, the U.S. Forest Service in Northern California let a contract to determine the effectiveness of canopy brush guards to provide operator protection from both falling snags and rollover.[5]

In 1958, the North Pacific Division of the U.S. Army Corps of Engineers initiated a design requirement for protective tractor canopies of sufficient strength to resist the forces of rollover on tractors used in heavy construction, and to provide a space for the operator in the event of upset so that he would not lose his life.[6]

This course of action was based upon an extensive safety research study conducted in the Portland (Oregon) District. Literature searches made during this study uncovered a wealth of historical background material. During the course of this three-month's study it was found that a D8 Caterpillar tractor used by a lumber company had been equipped for many years with a stout protective canopy constructed of four square posts made of two channel irons welded into box beams, and that this tractor, in 1947, while skidding logs on steep ground, rolled over sideways three times, yet the operator was unharmed and little mechanical damage resulted to the machine. In 1958 it was found that both operator and machine were still productive.

A summary of the Portland Districts Corps of Engineers' report was published in *Pacific Builder and Engineer*.[7] As a result of this study, the North Pacific Division of the Corps of Engineers adopted this requirement for rollover protective canopies on all tractors owned by either the government or contractors and used in the Division.[8] This specification for rollover protection reads as follows:

"*TRACTOR CANOPIES: All crawler-type tractors shall be equipped with steel canopies for the protection of operators from the hazards of rollover. The design details shall be equal to those established by the Oregon State Industrial Accident Commission. Additionally, canopy frames shall be constructed of material as listed below, or the equivalent in strength thereof, with minimum requirements as follows:*

Gross Weight Including Attachment Without Canopy		Pipe Diameter and ASA Schedule
Under 28,000 lbs.	or	2½ in.—Sched. 80
		2 in.—Sched. 160
28,000 to 58,000 lbs.		3 in.—Sched. 80
Over 58,000 lbs.		4 in.—Sched. 80

"*(1) All canopy frame shall be haunch braced to top support members at the intermediate post on each side of the frame with one transverse and two longitudinal braces twelve inches long. Braces shall be of the same diameter ASA Schedule pipe as shown above for the various weight tractors or shall be gusset braces of equivalent strength.*

"(2) All canopy mountings or attaching brackets shall be constructed and secured to the tractor in a manner to provide support equal to the structural strength of the upright members of the canopy."

This standard was an inspection-guide type of specification that allowed field personnel to determine by measurement whether the canopy could resist "W2" forces (required loadings twice the weight of the machine from a lateral and vertical impact). Some 1,000 tractors ranging in age of manufacture back to the late 1930's were successfully retrofitted.

Angle brackets with gussets were bolted to the tractor frame. These angle brackets supported the rollover protective canopy. The canopy legs were usually attached to the brackets by a well and pin-type connection. In the event of rollover, minor distortion (absorption of energy) of the brackets would occur, preventing damage to the main tractor frame, while effectively anchoring and supporting the protective canopy.

In over 90% of the reported rollover accidents, the tractor harmlessly rolled over 90° onto its side and came to rest, avoiding death or serious injury to the operator. Several cases of tractors rolling off steep bluffs were recorded during this study in which the operator escaped with only a few scrapes and bruises.

Acceptance and development of new standards

Many districts outside the North Pacific Division voluntarily implemented the new standard. By 1959 the effectiveness of rollover protective structures was an accepted fact within most of the Corps of Engineers' civil construction districts. In February of 1959, *Western Construction News* predicted that rollover protection was a coming trend.[9]

The State of Oregon's Industrial Accident Commission, upgraded its brush guard canopy requirements to incorporate the Portland District rollover protection Standards.

The Bureau of Reclamation, U.S. Department of the Interior, also adopted the Corps' standard for rollover protection.

Early in the 1960's the Operating Engineers' Union requested that the State of California adopt a rollover protective standard.

In the mid-1960's manufacturers were confronted with a variety of requirements for rollover and operator protection. The industry's response was to assign committees within the Society of Automotive Engineers (SAE) and the American Society of Agricultural Engineers (ASAE) to investigate the development of industry design criteria for Rollover Protective Systems (ROPS).

In 1965 the International Labour Office, Geneva, Switzerland, in its manual, *Safety & Health in Agricultural Work,* required the following:

"437. Tractor cabs or frames should be of sufficient strength and be adequately fixed to the tractor so as to provide satisfactory protection for the driver and passengers inside the cab in case the tractor overturns sideways or backwards."

With pending publication of SAE and ASAE standards, the National Safety Council, on January 26, 1967, published a "Resolution on Overturn Protection for Farm Tractor Operators" which recommended that all manufacturers install rollover protective systems as standard equipment. This document was developed and approved by the Farm Conference of the National Safety Council and stated:

"Farm tractor overturn accidents result from an adverse interaction of the operator, tractor and environment, and are known to claim more than 500 lives each year.

"It is recognized that persons when operating farm tractors should exercise reasonable care and adhere to recommended safety practices. It is further recognized that operator error cannot be totally controlled; thus, tractor overturn accidents, and the resultant deaths and injuries, are likely to continue.

"Considerable evidence is available to show that protective frames and crush-resistant cabs have potential to sharply reduce the number and severity of injuries to operators involved in tractor overturns.

"The Farm Conference therefore urges action on the following recommendations:

1. That the American Society of Agricultural Engineers and the Society of Automotive Engineers adopt performance standards for basic overturn protection on farm tractors, including protective frames and crush-resistant cabs.

2. That the farm equipment industry make available, as standard equipment, basic operator overturn protection on farm tractors that will conform to American Society of Agricultural Engineers' and Society of Automotive Engineers' standards."

The manufacturing industry *never* fulfilled this mandate, and the safety of farm tractor operators was recklessly disregarded. This is a clear indication that this industry does not have an authoritative and functioning safety program at corporate level, with authority over all facets of safety, including not only employee safety, but product safety. And it also appears that the Boards of Directors of these manufacturers do not actively take interest in or control necessary decision-making regarding product safety.

By November, 1967, SAE approved a "Minimum Performance

> *". . . protective frames and crush-resistant cabs have potential to sharply reduce the number . . . of injuries . . ."*

Criteria for Roll-Over Protective Structures for Rubber-Tired, Self-Propelled Scrapers," SAE J320. In April, 1968, SAE J334, "Protective Frame Test Procedures and Performance Requirements," was adopted to provide operator protection from rollover for both light industrial and agricultural tractors. SAE J396 for graders, SAE J395 for crawler tractors and loaders, and SAE J394 for rubber-tired front-end loaders and rubber-tired dozers were adopted in 1969.

In August, 1967, ASAE published its "Operator Protection for Wheel-Type Agricultural Tractors," S305, and "Protective Frame for Agricultural Tractors-Test Procedures and Performance Requirements," S306. These two standards were consistent with SAE rollover requirements.

The test procedures incorporated into these standards were patterned after European test concepts previously developed in the late 1950's and early 1960's. References 10 to 15 are some of the available European test reports on ROPS.

Construction equipment requirements

In 1969 there was quick agreement among newly appointed members of the U.S. Secretary of Labor's Construction Safety Advisory Committee that rollover protection was needed nationwide on all tractors used on federally funded construction. There was no opposition expressed by contractors to this provision, since they knew the cost of ROPS would be included as a bid item.

When the Occupational Safety & Health Act (OSHA) was enacted in 1972, general industry standards would put the burden on all employers to safeguard their tractors with ROPS. At this point a lot of short term economics came into play, and after much debate, only construction equipment back to 1969 was required to be retrofitted with ROPS.

After OSHA hearings, the requirement for installation of ROPS on tractors in agriculture was initially excluded. As earlier shown, the Corps of Engineers was able to make all tractors used on construction projects safer, but OSHA standards, as published, only assured for the safety of the operator by including a requirement for ROPS on machines made after 1969.

Concern for the lives and safety of tractor operators took a back seat when a U.S. Department of Labor study was contracted to determine the engineering and economic feasibility of retrofitting ROPS on pre-July 1, 1969, construction equipment. The final report, published on July 15, 1974, resulted in a policy which became a "license to kill," because it concluded that it was not economically feasible to provide ROPS on equipment older than 1969.[16] The stage was now set to raise the issue of tractor safety in courts across the land, and this has cost tractor companies several million dollars for jury awards to some of the widows, together with the cost of legal services defending such cases.

Agricultural tractor protection

In 1975, I received the following reply from Norman C. Mindrum, Director, National 4-H Service Committee, to a letter I had written co-addressed to the President of the American Society of Agricultural Engineers and the President of the National Safety Council, recommending the use of rollover protective structures on tractors used during the 4-H Tractor Contests:

"As former Vice President for Farms of the National Safety Council at the time of the Tractor Overturn Prevention and Protective Campaign, I recognize the safety provided by the roll-over protective structures (ROPS). Please be assured that our staff representative working with the Extension Service in planning the regional tractor contests will recommend that every effort be made to obtain tractors equipped with these safety structures.

"Although ROPS are available on newer tractors and will be mandated under OSHA for employee-driven tractors manufactured after October 25, 1976, tractors equipped with ROPS have been difficult to obtain for contests in the past. The Extension Service conducting these events must rely on the generosity of manufacturers, dealers or, in some cases, on local farmers for loan of the equipment.

"We recognize educational value in having ROPS-equipped tractors as well as in providing a safe environment for the contestants at Tractor Contests. In this connection, I want to point out that strong emphasis has always been placed on selecting contest sites which are level and free from obstacles. Also, it should be noted that the tractors used in the contests do not operate under a load condition and do not travel at speeds exceeding 3-4 miles per hour.

"You may be assured that we will continue to place strong emphasis on safety at the 4-H Tractor events."

> ... deaths and injuries would have been prevented had standard protective canopies ... been included ...

It appears that voluntary programs for our youth do not develop a "go" or "no go" priority for safety. In aviation, if necessary safeguards are not available, the plane is not flown. To me, it is poor judgment to develop within our youth the idea that if the safeguard is unavailable, we can go ahead and assume the risk that the hazardous equipment creates, rather than insist upon proper safeguards.

"Rollover Protective Structures for Farm and Construction Tractors—a 50 Year Review," SAE paper #710508,[17] gives insight into the frequency of rollover for the period from 1920 to 1970. It states that during this fifty-year period, some 30,000 tractor rollovers occurred, primarily in farming. Using this figure as a base, it can be estimated that a total of some 50,000 tractor rollovers occurred in farming and all other industries (construction, logging, mining, materials handling, etc), and that these occurrences were widely dispersed geographically among the various industries.

From this data, it could have been easily projected by tractor manufacturers in 1970 that some 1,000 deaths per year from tractor rollover would occur. When severely crippling injuries from related hazards (being pitched off or displaced by falling, swinging, or intruding objects) were added to this projection, the estimate was raised to approximately 10,000 accidents per year. This is a casualty rate comparable only to the losses sustained in a war.

Most of these deaths and injuries would have been prevented had standard protective canopies strong enough to resist the forces of rollover been included on all tractors manufactured. Our medical scientists would never have ignored for so long an epidemiology of this magnitude for any disease. This attitude on the part of manufacturers shows a lack of concern and a disregard for the safety of the buyers and users of their products.

Feasibility of design

The "state of the art" at the time the first tractor was manufactured made it entirely feasible to design a canopy for a tractor to give protection from all foreseeable hazards inherent in known circumstances of tractor use. In fact, early steam tractors included substantial steel cabs for the protection of the operator, and examination of some thirty U.S. patents shows that the know-how to design tractor operator protection was clearly within the scope of our technology during the early days of tractor production.

Also, since the early 1950's an ample number of suppliers have successfully designed, manufactured and sold protective structures advertised as being able to resist the forces of upset or complete rollover. Add to this the fact that over three hundred articles have been published during the past sixty years that addressed the hazards of rollover and the need for overhead protection. It is very apparent that the "state of the art" concepts had been developed but not initiated by manufacturers early in tractor history.

The Canadians, Swedish, New Zealanders, and English have been acutely aware of the hazard of rollover, and in the early fifties were conducting tests as a basis for safety requirements. By the mid-1960's many of these countries required that acceptable ROPS be included on all U.S.-manufactured tractors imported by them. However, on tractors exported to underdeveloped countries, no steps were voluntarily taken by U.S. manufacturers to include ROPS as standard equipment.

By not incorporating available safeguards into design, manufacturers were put in a position of selling machines which were unreasonably dangerous. An unreasonably dangerous machine is one which will foreseeably kill or maim, because reasonable safeguards are not included in design, and at time of manufacture to protect the operator from inherent hazards in the foreseeable use of the machine.

Surely the addition of a simple, substantial, steel protective frame was within design and manufacturing capabilities of the 1920's when gasoline and diesel tractors were first being developed and the inherent hazards were beginning to be noted. It is, indeed, a sad indictment upon our industrial age that the safety of the user/operator was totally neglected in design and manufacture of these machines.

This observation was clearly stated in Allen Toffler's book, *"The Third Wave,"* in which it is stated that industrial plants that damage the environment may not be as efficient and cost-productive as we think. This same philosophy can be applied to a machine which kills and maims—it, too, is not truly productive.

The history of ROPS gives good insight into the reluctance which has seemed to prevail among some manufacturers of equipment to incorporate necessary safeguards in design so that users of equipment will be protected from hazards encountered in the circumstances of a machine's use. It appears that safety of the tractor operator was not one of high priority.

Duty and responsibility

It appears that although designers and manufacturers of tractors have made in-depth studies to analyze the sales market for tractors, they have done very little investigation into operator death and injury resulting from the use of these machines. Manufacturers have a duty to familiarize themselves with accident occurrences associated with use of their equipment, so design improvements can be incorporated to eliminate or minimize the risk of death or injury resulting from a machine's normal foreseeable use.

It also seems that management often neglects or avoids delegating safety responsibility in corporate policy. Management should assure that safety is not compromised by other divisions in the organization. Success in the area of product safety is usually dependent upon a management structure which provides for a corporate safety director who is directly responsible to the chief corporate executive officer and who provides as-

Management should assure that safety is not compromised...

sistance to the engineering staff.

In order to assure that a safe and productive product is entered into the market, such a corporate safety director must have the authority to require that the accident experience of each piece of equipment be researched, that appropriate literature searches be made, and that the latest in safety technology is made available to designers so that inherent hazards are eliminated or minimized through design safety improvement. Additionally, management must assure that the sales staff informs the using public about any inherent hazards and how any safeguards provided will minimize the risk of death or injury while using the equipment.

Operator hazards

The ability, versatility, traction, and power of both crawler and wheel tractors, front-end loaders and dozers to manuever on steep slopes and rough terrain should not be compromised by sole reliance upon perfect operator performance to avoid circumstances which are conducive to rollover or falling objects. If work assignments for such machines were to preclude use when such dangers were encountered, the machine's utility would be vastly reduced.

To compensate for foreseeable variables of human performance, judgment, or failure to recognize the potential danger posed by the types of work these tractors are designed to do, and because of the high-risk of death or severe injury to which a tractor operator is exposed, manufacturers should learn that they have four basic avenues to pursue to control hazards which endanger operators or users:

1. Incorporate safeguards in initial design criteria to assure for the safety of the operator or users.
2. Sell only equipment with safeguards as standard components of the machine.
3. When field safeguards are developed by others, by specialty suppliers, or by competing manufacturers, initiate the following retrofit programs:
    a. *Equipment bought or taken in on trade by the manufacturer or its agents:* Equip with safeguards before resale.
    b. *Equipment repaired or serviced by the manufacturer or its agents:* Equip with safeguards when machines are brought in for repair or service in a special program for owners.
    c. *Equipment not repaired or serviced by the manufacturer:* Develop an information program for owners of its product telling of the need to equip the dangerous machine with a safeguard to avoid death or serious injury.
4. Develop a comprehensive public awareness program promoting the need and use of protective safeguards to acquaint potential purchasers and users with the seriousness of the hazard.
5. Every piece of equipment needs to carry with it explicit safeguard warnings, starting with sales brochures and point-of-operation labeling through a comprehensive discussion of the hazard in the operator's and maintenance manuals. A mere adhesive caution label is insufficient. Warning labels must follow basic DANGER format. Additionally, the operator's manual

should be carried in a watertight container so that every user has the necessary safety information available. This container should have a label giving a phone number for replacement of a lost manual.

Legal and monetary problems

Since the early 1970's, many product liability lawsuits have been filed because of death or injury caused by failure to provide ROPS. This number grows each year. A close examination of the tractor rollover casualty rate shows how the industry's election to let tractor operators assume the risk of death, rather than give them a safe machine, has been translated into the filing of numerous lawsuits.

By avoiding a logical installation of a much needed safeguard or developing a retrofit program for ROPS, the industry has defaulted its safety responsibility. Conservatively speaking, these cases, whether or not successfully achieving recovery for the injured, have resulted in losses to manufacturers of many millions of dollars.

Recently the United States Court of Appeals for the 3rd Circuit in Philadelphia, Pennsylvania, No. 81-2720, in Hammond v. International Harvester, upheld a plaintiff's verdict and established that the purchaser's deletion of the safeguard from the order was not a defense. This verdict also established that the argument that the machine was older than the effective date of the OSHA requirement for ROPS was no defense for a manufacturer, because of the seriousness of the consequence of the hazard.

This is not really new law since same concepts were written four to six thousand years ago in the *Bible*. In the "Good News" Edition, Exodus, Chapter 21, Verses 29 and 30, read:

> "But if the bull had been in the habit of attacking people and its owner had been warned, but did not keep it penned up—then if it gores someone to death, it is to be stoned, and its owner is to be put to death also. However, if the owner is allowed to pay a fine to save his life, he must pay the full amount required."

We can only conclude that if a manufacturer is aware that a machine is dangerous because of repetitious litigation arising from the same hazard, that it has a clear duty to safeguard the machine by im-

...a manufacturer... has a clear duty to safeguard the machine by improving the design...

proving the design on future models and retrofitting those already manufactured. I believe it would be less expensive for tractor manufacturers to retrofit all their tractors with ROPS rather than continue to face the cost of future liability, not even considering the human suffering which could be prevented.

Unexpected death or severe injury not only disrupts family relationships but often creates claims upon the Social Security Fund over and above worker compensation claims due the family. This places an unfair economic burden upon all taxpayers, who must then pay for the negligence of manufacturers who do not face up to their safety responsibilities. No wonder both the Social Security Fund and many of these manufacturers are in monetary straits.

Industry's neglect of hazards which can be easily eliminated in safe design has defrauded everyone:

a. The operator—by giving him an unsafe machine that endangers his life.
b. The family of the deceased or injured operator—by unnecessarily depriving them of love and economic support.
c. The community—by causing a drain upon welfare and social service funds to provide for rehabilitation and family care.
d. The employer—by increasing the cost of worker compensation.
e. The U.S. Social Security Fund—by transferring responsibility to this ailing fund.
f. The stockholders and investors—by the sale of an unsafe and dangerous product, exposing the corporation to lawsuits and diminishing the very profitability of the corporation.

If the total cost of a death for lack of a safeguard was to be estimated at $2,000,000 for each occurrence when including all legal costs and loss to the community, for every one hundred wrongful death claims the estimated economic impact would be $200,000,000. For a fraction of this cost, every existing machine unequipped with a safeguard could be retrofitted. Production of a safe product with safeguards as standard equipment would avoid the huge monetary outlay for damages which has been created by an unsound corporate position on safety.

Because of the fear that any standard installation of a safeguard or retrofit program might imply guilt, corporate legal counsel continue to advise management to defend every claim without considering settlement. This approach does not reflect any of our country's basic ethical concepts that embody a strong sense of care for our fellow beings' safety and welfare.

Since World War II, we have witnessed the intense and increasing dissatisfaction of our youth with rigid and unaccommodating management practices that do not apply available technology for the protection and benefit of the public. We have ample safety technology that could provide substantial savings in life and money compared to the burden foisted upon our unsuspecting, benevolent community for unnecessary wrongful deaths and maiming injuries.

James E. Lardner, Vice President of Deere & Company, in the August 23, 1982, issue of *Industry Week*, explains why he feels U.S. manufacturers are missing out when it comes to new technology: the leaders are too old, inflexible, and do not understand the new technology; and the younger "middle managers" who do understand do not have the authority to make changes. This can be applied to safety as well as all other aspects of equipment design and manufacture. This lack of understanding on the importance of design safety is one of the causes which has led many manufacturers to the brink of bankruptcy.

Safety training, safe design

Unless safety training and safe design go hand in hand, the loss of human life and economic hardship will continue to be felt by both individuals and corporations. Injuries and death are best reduced by eliminating, isolating, or guarding hazards. This is the great lesson the history of ROPS has taught. Twenty five years is too long a time to rely on operator skill alone to accomplish the desired result. System safety analysis of all products at time of design is needed.

Accident statistics are also an important component of system safety—to monitor the effectiveness of design. For twenty five years rollover protection has shown a dramatic drop in deaths on tractors equipped with ROPS. The older models of industrial tractors and farm tractors unequipped with ROPS are still causing nearly 100 deaths a year. Fortunately, many of the older models are being equipped with ROPS on a voluntary basis by perceptive owners.

Using the projection figures stated earlier, with an even larger and growing population of tractors, perhaps over 1,000 lives a year are now saved as newer tractors equipped with ROPS vastly outnumber the older ones not so equipped. The risk of death and injury has lessened and would nearly vanish if all farm tractors and older construction models were to be retrofitted, just as death from smallpox has been eliminated by vaccination.

Those who do not assure for safety in design are only planting the seeds of corporate destruction and public distrust. Safety professionals have a leading role among the decision-makers of management, but their success will depend upon the development of design remedies to overcome hazards rather than relying solely upon equipment operators to achieve accident free performance.

The safety professional's role of identifying the hazard, quantifying the risk, communicating the resulting danger in terms of actual peril, and developing safeguards is second to none. The doctor heals; the lawyer recovers damages; the insurer uses the insured's money to transfer the risk; but the safety professional can change the course of events and prevent predictable accidents from occurring, thus saving lives and property and reducing costs.

No excuses

The excuse that there are no clearcut governmental regulations for installation of safety devices such as ROPS for farm machinery and older construction equipment is not valid. Do we have to have more regulations to live up to our moral responsibilities when it comes to the lives of those who must use our products?

Industry is constantly complaining about too much government regulation. The only way to eliminate such regulation is to provide necessary safeguards to control hazards. No manufacturer should wait until regulated to include safeguards on any product. This only brings on the "House That Jack Built" syndrome:

This is the hazard which caused an accident;
This is the accident which caused a lawsuit;
This is the lawsuit which brought about a new safety law;
This is the safety law which authorized more government control;
This is the governmental control which upset management;
This is upset management which misspent its time complaining about governmental control;
This is misspent time which could have been used to look for hazards;
This is the hazard . . .

. . . safety in design carries . . . rewards in number of lives saved . . .

What have we learned from twenty five years of ROPS? I hope we have learned that a priority for safety in design carries with it impressive rewards in number of lives saved. No one should have to be trained to avoid inherent hazards in a product when safe products are achievable. Let's not wait sixty years after a product is introduced into the marketplace to make it safe. The toll is too high!

References

1. Edward R. Hewitt, *SAE Transactions,* Vol. 14, Part I, pg. 83+, "The Principles of the Wheeled Farm Tractor", 1919.
2. E. G. McKibben, *Research in Agricultural Engineering,* "Kinematics and Dynamics of the Wheel Type Farm Tractor", February, April, May, June, July 1927.
3. a. Logging & Sawmill Safety Orders, California Administrative Code April 20, 1945
   b. General Safety Requirements, EM 385-1-1, War Dept., Corps of Engineers, January 1946
   c. Logging and Sawmill Safety Code, Oregon, Industrial Accident Commission, January 8, 1948
   d. Safety Standards for Logging Operations, State of Washington, 1948
4. University of California Agricultural Extension Service, "Driver Safety Frame", December 1957.
5. Donald E. Platz, E&R Development Company, Report No. G22-2, "Tractor Canopy Impact Tests Conducted at Redding California During June 1956", July 20, 1956.
6. U.S. Army Engineer District, Portland, Corps of Engineers, Safety Branch, "Evaluation of Tractor Canopies in Rollover Accidents"
7. David V. MacCollum, *Pacific Builder & Engineer,* "Tractor Canopies," October 1958.
8. U.S. Army Engineer Division, North Pacific, Corps of Engineers, "General Safety Requirements EM 385-1-1, 13 March 1958," July 15, 1960.
9. *Western Construction,* "Tractor Canopies—Start of a trend?," February 1959.
10. National Swedish Testing Inst, "Tractor Safety Cabs, Test Methods and Experience Gained, During Ordinary Farm Work in Sweden," 1962.
11. New Zealand Agricultural Engineering Inst., Lincoln College, "Public Test Report No. T/I, Bonser Safety Frame for David Brown Tractors 850, 880, 900, 950 and 990," August 1965.
12. Edmund J. Zeglen, ASAE 59th Annual Meeting, Amherst, MA, Paper #66-111, "New Tractor Development at Massey-Ferguson for 1966," June 26, 1966.
13. New Zealand Agricultural Engineering Inst., Lincoln College, "Public Test Report No. T/3, Fergtrac Safety Frame for Ferguson TEA-2-Tractors," September 1966.
14. Department of Labour, New Zealand, "Bush Tractor Canopies," 1967.
15. E. M. Watson, New Zealand Agricultural Engineering Inst., Research Publ. R/1, "The Structural Testing of Tractor Safety Frames," May 1967.
16. Occupational Safety & Health Administration, U.S. Department of Labor, DOL Contract No. L-73-158, "Study to Determine the Engineering and Economic Feasibility of Retrofitting ROPS on Pre-July 1, 1969 Construction Equipment, VOL II, Final Report," July 15, 1974.
17. James F. Arndt, Deere & Company, SAE Earthmoving Industry Conference, "Roll-Over Protective Structures for Farm & Construction Tractors—A 50-Year Review," April 5-7, 1971.

Hazard control of liquid oxygen systems

Artist's conception of a flame-seared liquid nitrogen tank standing alone following a liquid oxygen tank rupture and blastoff. Molten aluminum and oxygen was sprayed on ground below the oxygen tank as it lifted into the air. A 9000 gallon aluminum LOX tank stood beside the 9000 gallon, 40 foot high, 10 foot diameter liquid nitrogen tank before the LOX tank violently separated circumferentially from its bottom and took off into the air like a rocket with a loud boom. Some described a rocket type tail flame and a flight several hundred feet high followed by the tank falling to earth over 300 feet away. The leaning "structures" depicted on the left side of the 40 foot high tank are a series of many oxygen gasifiers. On the right are many gasifiers which were ripped loose and blown aside.

by William W. Allison

Liquid oxygen (LOX) can be used safely if the system and facility is properly designed, cleaned and maintained; and if all personnel are specifically and adequately trained. Ignition, explosive reactions, low temperature effects on materials and personnel are among many peculiar characteristics that must be adequately considered.

Design considerations are briefly covered and are fairly specific in the literature. However, cleaning requirements vary from requiring zero parts per million of any foreign materials to the very vague requirement of "making sure it is clean." Consequently a composite of the best practicable cleaning requirements was composed, utilizing experience in the control of hazards as well as the expertise and experiences of others.

Cryogenics exhibit actions and reactions that demand knowledge, skill and caution to utilize their benefits. In addition to the better known hazards, the use of large quantities and/or high flow rates create problems beyond those encountered in low quantity and/or low flow rate commercial applications. Sudden changes in flow rates from closed to wide open or wide open to closed can cause serious accidents.

→

Personnel contact hazards
Neither the extremely cold liquid nor the metal that holds it or is cooled by it should be touched. Severe cold burns and freezing a finger to icicle brittleness can result.

Detonation and fire hazards
Liquid oxygen (LOX) presents hazards peculiar to its liquid state. Liquid oxygen can detonate when a hammer or other heavy object is dropped on a spill of LOX. It makes clothing and asbestos gloves burn like a dry straw broom torch. It explodes or burns furiously on contact with even a drop of grease or oil in a pipeline, a regulator or gauge.

Liquid to gas pressure hazards
In piping, vaporizing or collecting any cryogenic, a relief valve must be installed between each shutoff valve, regulator or flow restrictor to prevent the rapid over pressurization which results as the liquid changes to the gas phase. Confinement of cryogenic fluids can readily result in extremely high pressures beyond the strength of piping or vessels. Liquid to gas expansion ratios are 862 for oxygen, 847 for nitrogen, 726 for air, and 85 for hydrogen (B.P. to 70°F). If 40 pounds of liquid oxygen were placed in a one cubic foot stainless steel cylinder, a void space of 34 percent of the volume would remain. If this cylinder were closed and allowed to warm up to room temperature it would attain a pressure of 10,000 pounds per square inch.

Reaction hazards
Reactions occur due to contact of oxygen with oil, grease, contaminants, asphalt and most organic materials. Concrete rather than asphalt must be used for unloading pads, curbing, etc. Cleaning to remove all contaminants, weld slag and all foreign materials is mandatory. See section on cleaning requirements.

Spots of weld, slugs of insolubles, rapid valve opening and gas bubbles are among the causes of explosions and detonations in liquid oxygen systems. Minute contamination in the order of several hundred parts per million of acetylene or hydrocarbons can lead to explosions or detonations. Regular monitoring should be done for contaminants of minute quantities such as acetylene and hydrocarbons.

Positive check valves and other adequate means are required to prevent oxygen backflow into parts of the system that may not be super-clean or free of noncompatible reactants.

Brittle fracture and coefficients of expansion
Brittle failure of metals at cryogenic temperatures must be avoided by proper selection of materials. Carbon steels for example have precipitous transition to low breaking strength impact values. The sharply varying differential expansion or contraction of materials from temperature changes several hundred degrees in magnitude must not be overlooked.

Missiles and pipe whipping hazards
Missiles and pipe whip can result from the sudden release of the confined gas from cryogenic fluid warming or from compressed gas containers. The common oxygen cylinder weighing 140 pounds containing 2500 psi could attain a velocity of nearly 50 feet per second if a valve were broken off the cylinder. Obviously strong tiedown means are required to restrain cylinders, piping and vessels.

Other considerations
A reinforced concrete barrier to prevent supply trucks or other vehicles from striking the piping, fixtures or storage tank is recommended. A fire hose hydrant, hose and/or a deluge spray monitor, depending on quantities of oxygen, should be located nearby to cool vessels and piping and to wash/evaporate a leak or spill.

The science and art of cryogenic hazard control have advanced considerably since the liquid gas catastrophe in Ohio in the early forties. Nevertheless, not long ago an aluminum 9000 gallon LOX tank ruptured. The tank was reportedly propelled over 300 feet. The shock wave inflicted heavy damage to buildings with steel siding and injured personnel. Oxygen entered damaged buildings nearly 100 feet away and fires ensued due to the oxygen enriched atmospheres.

Serious equipment, tank, and pipe ruptures and fires, as well as enriched oxygen atmosphere fires and fatalities have occurred in the space program. High oxygen flow rates and/or large volume usage dictates the need for increased safety requirements. The potential risk is more severe and requires more extensive hazard control measures than commercial low flow, low volume usage.

Serious accidents have occurred almost simultaneously with the quick opening of shutoff valves against high pressure oxygen. Compression wave front temperatures that can ignite steel can occur. The presence of sand, scale, welding bead, CO_2, any particulate, grease or oil can conceivably produce ignition, yet analysis of the liquid could show less than one part per million of hydrocarbons.

Oxygen and organic foams can be shock sensitive and must not be used in or near liquid oxygen systems. Rupture discs must have readable burst pressure/temperature markings permanently affixed to the discs.

Distance and either masonry or reinforced concrete buildings with no windows are indicated for occupied buildings in areas where large quantities of oxygen are stored or high flow rates are used or oxygen is manufactured. Corrugated asbestos-cement, steel or aluminum siding fails at one to two pounds of blast overpressure. Cinder block fails at two to three, while eight to twelve inch brick walls fail at seven to eight psi of blast or shock wave overpressure.

Oxygen vapor tends to flow downhill. Provision should be made to confine spills within diked areas and to avoid any possibility of oxygen liquid or vapor from flowing downhill to areas where organic materials or personnel clothing may be saturated with oxygen. By the same token, oil or other liquid spills must not be permitted to flow into an

> *The shock wave inflicted heavy damage to buildings with steel siding and injured personnel.*

oxygen storage, compressor or spill area. These problems must be considered in plant layout of equipment and traffic and pipeline arrangements.

By analyzing separately for acetylene and providing continuous total hydrocarbon analysis at oxygen manufacturing plants in the 1950's, the safety of such plants was greatly increased. However, the use of large quantities of oxygen at missile sites introduced increased opportunities for buildup of concentration of contaminants. Stricter cleanliness, contaminant analyses and care in preventing mechanical and dynamic high flow rate induced problems became evident.

Finally, and perhaps most important, scrupulous selection of suppliers and engineers, knowledgeable personnel, training, safe operating procedures, written work permits, emergency procedures and operating logs are essential.

Material selection and oxygen system cleaning requirements

Cleaning and cleanliness, in the ordinary sense, is not a sufficient or safe criterion when dealing with liquid oxygen or gas. The extremely reactive nature of materials in contact with liquid oxygen prohibits the presence of liquids and liquid residues of over milligram quantities and the presence of solids over a specified micron size.

Oxygen systems must be free of cloth lint, hair, brush bristles, wire bristles, lint or oil from gloves, skin, tools or equipment. Oxygen systems must also be free of mercury, mercury instruments, plastics, except for those specified below, hydrocarbons, metallic oxides, chips, filings, tar asphalts, weld slag, paints, and petroleum based lubricants.

Some of the materials that are known to react with violence approaching a detonation are oil, grease, asphalt, kerosene, cloth, wood, tar, and paint. Even metal is potential fuel in an oxygen system. Only the sparking of the scale from the inner walls of carbon steel, for example, or the heat from sudden compression or shock waves from rapidly opening a large valve can ignite the piping or valving and dump the system with potentially catastrophic consequences.

Problems

Problems that have been encountered in cleaning include:

1. During use of vacuum pumps without adequate cold traps, there was back-diffusion of vacuum pump lubricant into oxygen vessels from normal operation of vacuum pumps and pump lubricant carried back into the liquid oxygen container when the pump stopped.
2. Analysis of a liquid oxygen vessel explosion indicated that hydrocarbon cleaning solvent was trapped in a small well formed by an interior reinforcing bar and the vessel wall.
3. Phosphate and silicate cleaning agents have precipitated on the interior walls. Low temperatures of the cleaning solution, greases, non-volatile hydrocarbon residues, vacuum pump lubricating oils and leaving a cleaning solution stand are among the causes of precipitation.
4. Use of wood plugs and rags have contaminated cleaned vessels with wood splinters, lint and threads.
5. Chlorothene from a supplier was found (during lab tests before use) to leave an oily non-volatile residue in the evaporation dish. This was caused by presence of moisture. In the presence of moisture, chlorothene is corrosive and moisture destroys the inhibitor.
6. Chlorothene (1, 1, 1 trichloroethane) methyl chloroform with added inhibitor has no known flash point in air, but it is flammable and explosive in oxygen as are all hydrocarbons under certain (generally unknown) conditions.
7. Oil contamination from drying a clean, rinsed tank with air from an oil lubricated compressor.

Artist's conception of major portion of 40 foot liquid oxygen tank after it landed over 300 feet away. Outer tank shell was partially burned off. *Questions include:* Did the safety relief valve freeze and permit overpressurization? Did the overpressure then separate the bottom head from the main portion of the aluminum pressure vessel and thus expose virgin aluminum to liquid oxygen? Did the nonoxidized aluminum then fuel the oxygen and send the LOX tank and its outer shell into the air? Were contaminants involved in maintenance, in the delivery tank truck, or in hooking up for refilling the tank prior to the tank rupture? Did refilling result in heat absorption and did large quantity flow followed by shutdown shortly after refilling combine to cause a temperature/pressure increase and safety valve freezing?

Materials
The potentially violent reactivity of oxygen with many materials dictates both strict cleaning requirements and very careful selection of the proper materials for vessels, pipes, valves, other components, and lubricants. Because the materials used will affect the selection of cleaning agents and methods, the two subjects cannot be treated separately.

When selecting materials for liquid service, consideration should be given to physical properties at low temperature, and the reactivity of the material with liquid oxygen. The ability to withstand stress concentrations, particularly those resulting from sudden temperature changes, is important.

Metals
Metals to be used in liquid oxygen equipment should possess satisfactory physical properties at extremely low operating temperatures. Some metals react violently with liquid oxygen under certain conditions. Titanium is not permitted in liquid oxygen storage or test facility systems since its reaction with liquid oxygen may completely consume the metal. Carbon steel is extremely subject to brittle fracture at low temperatures. The carbon steels and steel alloys also present a problem in the high pressure, high velocity gas system of picking up oxide particles

from the pipe walls which strike elbows, etc. and ignite. Design criteria for carbon steel lines in gaseous oxygen service is a maximum pressure of 450 psig and a velocity restriction of 25 fps at 150 psig.*

Stainless steel or nonferrous sections, 3 feet long, must be used at elbows and at no more than 100 foot intervals, if carbon steel is used. The following metals are recommended for service with liquid oxygen:

1. Stainless steel types
   304   316   304L
   310   321   304ELC
2. 9 percent nickel steel alloy
3. Copper and copper alloys
   copper   aluminum
   naval brass   bronze
   admiralty   cupro-nickel
     brass
4. Aluminum and aluminum alloy types
   1000 3000 5083 5454 6062
   2014 5050 5085 5456 6063
   2024 5052 5154 6061 7075
5. Nickel and nickel alloys
   nickel   Inconel-X
   Rene 41   Hastelloy B

Non-metals
The number of acceptable non-metals is small due to the strong oxidizing properties of oxygen and the extremely low temperatures encountered. The following list contains the non-metals known to be acceptable:
1. tetrafluoroethylene Polymer (TFE, Halon TFE, Teflon, or equivalent).
2. unplasticized chlorotrifluoroethylene Polymer (Kel F, Halon CTF, or equivalent).
3. asbestos (in encapsulated forms).
4. special silicone rubbers (only if approved by the oxygen system vendor).

Lubricants
Liquid oxygen is a powerful oxidizing agent and no petroleum-based lubricant should be used. Special lubricants such as the fluorolubes or the perfluorocarbons are applicable.

Allowable velocities at other pressures may be determined by the following formula:

$$V \text{ (ft. per sec.)} = 100 \sqrt{\frac{165}{\text{psia}}}$$

Liquid oxygen is a powerful oxidizing agent —no petroleum-based lubricant should be used

---

Equipment

Containers
Liquid oxygen should be stored in stationary or mobile tanks of approved materials and construction. To ensure against defects in material or fabrication, the storage tanks should be tested as required by the provisions of applicable ASME specifications for unfired pressure vessels. (ASME Boiler and Pressure Vessel Code, Section VIII, American Society Mechanical Engineers.) Materials used for pressure vessels operating at temperatures less than $-20°F$ should be impact-tested in accordance with paragraph UG-84, Section VIII, of the ASME Boiler and Pressure Vessel Code. Storage containers should be vacuum-jacketed; the vacuum space may contain reflective insulation or powders. The storage tank itself should be of welded construction and should be equipped with an adequate pressure-relief system.

Pipes and fittings
Piping should conform to ANSI B31.1 Standard (American National Standard Institute). The pipes and fittings should be approved material and construction, and shall be hydrostatically tested at specified pressures. The use of welded and flanged connections whenever possible is recommended. Threaded connections sealed with litharge and water are permissible only when the other methods are not feasible. Fifty percent silver brazing is recommended for copper to copper, copper to stainless steel or to copper alloys. Soft solder is not recommended as it has little or no ductility at low temperatures.

The joining of dissimilar metals other than the above mentioned or substitution of metals requires metallurgical study and approval as structural failures can occur from different contraction or expansion from non-uniform temperature changes.

Gaskets
Gaskets may be made of soft metals selected from those listed above under "Metals." Organic material or flammable substances of any kind, including silicone rubbers except as specifically noted in "Non-metals" should not be allowed to come into contact with liquid oxygen.

Valves
The use of extended stem gate, globe, or ball valves provided with venting devices is recommended. Flat, optically ground seat and disc style safety relief valves with packed lift levers minimize icing and malfunction.

Pumps and Hose
Since the storage tanks may be designed with bottom outlets, flooded-suction centrifugal pumps may be used when gravity flow is not applicable. Only pumps and shaft seals designed for liquid oxygen service may be used. Details on these pumps and hoses may be secured from manufacturers of oxygen handling equipment. Hoses shall be of proper design and engineered specifically for liquid oxygen service.

Pressure Gauges
Liquid oxygen equipment shall be monitored with LOX-clean types of pressure gauges. In order to minimize operator reading errors, all pressure gauges used for a common purpose should have identical scales. Gauges shall be safety type with solid front and side cases with full blowout backs and either laminated polycarbonate or laminated safety glass dial covers.

Venting Systems and Safety Relief
The storage container itself shall be equipped with a bursting disc and a pressure-relief valve in parallel, both discharging to the outdoor atmosphere through an adequately sized vent line. The insulated area, between the inner and outer shells, should be equipped with either a rupture disc or a pressure-relief device, so that pressure cannot build up and rupture the vessel. All lines

and vessels in which liquid oxygen may be trapped between closed valves shall have pressure-relief valves; if it is likely that the relief valve may freeze, rupture discs shall also be provided.

Electrical Equipment
Electrical equipment containing arcing devices should be excluded from storage areas and transfer facilities. Oil bath electrical devices should not be used.

Cleaning Specifications
Each supplier of any and all equipment, vessels, piping, valves, gauges, and other components to be used in the oxygen system and all connections to the oxygen system up to the point where a block valve or other adequate isolation is provided must supply written specifications. Specifications shall cover cleaning, protection from recontamination and packaging. Specifications shall ensure the cleanliness and specify the microgram of liquid and micron size of solid particles remaining in the specific parts and system. Cleaning should be continued until the following requirements are met:

1. a system flush filter test indicates that all particles larger than 150 microns have been removed; and that all fibres longer than 200 microns have been removed.
2. total solids are less than 4 milligrams per square foot of critical surface area.
3. total hydrocarbons are less than 7 ppm by weight of the test fluid.
4. total non-volatile residue does not exceed 1 mg per square foot of critical surface areas.
5. total acetylene does not exceed 1 ppm of the test gas.

- Approved degreasers include the various non-hydrocarbon detergents such as trisodium phosphates and Oakites per manufacturer's recommendations and instructions.
- Cleaning of aluminum and plastics with hydrocarbon solvents is prohibited as they can alter the dimensions and properties of plastics and attack aluminum. Use only appropriate Turco, Oakite, or Diversey water soluble agents prescribed by the manufacturer for oxygen systems cleaning of the specific material(s) being cleaned.
- Filtered clean water rinses must follow the water soluble cleaning agent wash.
- Dry, oil free air drying by use of exhaust air movers and filtered, heated inlet air must follow the filtered clean water rinses.
- Aqueous ultrasonic cleaning is approved and the use of refrigerant 113 is also approved.
- If hydrocarbons are used for degreasing, they must be followed first by water soluble washing and finally by filtered clean water rinse.
- The use of certain stainless steels can preclude the necessity of acid pickling to remove rust scale. Acid pickling is usually required on non-stainless steel.
- The use of tapes to seal open tubing can leave an organic residue that reacts with oxygen; and the use of Teflon tape on threads frequently leads to tape pieces being carried into the stream and lodging in instruments, etc.

Final Check of Entire System
After all the cleaning steps and tests for cleanliness outlined above under Cleaning are completed and double checked, the entire system should be blown down with dry air followed by a liquid nitrogen flush. This cold shock will free particles such as slag which may be stuck to the metal. The liquid nitrogen flush should be followed by a thorough check and inspection for leaks in the system. The repairs of leaks should be followed by repeated tests with liquid nitrogen until there are no leaks in the system. This usually requires several days of inspection and repairs.

Special Oxygen Clean/Assembly/Work Area
1. An area must be clearly defined and controlled for the final cleaning, assembly and leak repairs. It must be kept free of combustibles in a clean and oil-free condition.
2. Work benches must be non-combustible and kept clean and free of grease, oil, etc.
3. Tools must be clean tools used only for oxygen system erection and assembly. Tools used for general work in other areas must not be permitted.
4. Personnel must clean their hands thoroughly and change into approved protective clothing. Coveralls, gloves, shoes and hands and hair must be free of grease, oil and other contaminants. No protective hand creams should be permitted. Lint free gloves are required. Paper wiping materials rather than rags should be used—clean-lint-free cloth may be used.

References
1. NFPA Nos. 50 and 53 M National Fire Protection Association, 470 Atlantic Avenue, Boston.
2. "Hazards of Chemicals, Rockets and Propellants Handbook, Vol. III, Liquid Propellant Handling, Storage and Transportation," DOD-AD-870259-CPIA/194-May 1972.
3. "Equipment Cleaned for Oxygen Service," Pamphlet G-4.1, Compressed Gas Association Inc., New York, New York.
4. Gaseous Oxygen Data Sheet 472, National Safety Council, Chicago, Ill.
5. "Oxygen System Cleaning Requirements," Cryogenics and Industrial Gasses, Nov./Dec. 1973 and March/April 1974.
6. Zabetakis, Michael G., *Safety With Cryogenic Fluids,* Plenum Press, New York, 1967.
7. *Proceedings 1959 Cryogenic Safety Conference,* Sponsored by Air Products Inc., July 1959.
8. *Cryogenics Safety Manual,* British Cryogenics Council, London, 1970.

There are five requirements to be met before cleaning equipment and components.

An addenda
Hazard control of oxygen systems

by William W. Allison

The original paper, "Hazard Prevention of Liquid Oxygen (LOX) Systems," *Professional Safety* Jan. 1979, was intended to address only LOX systems, emphasizing cleaning requirements and compatible materials. However, gaseous oxygen is both a product and by product of LOX and thus a part of LOX systems. Also additional data has become available since the paper was developed (long before printing).

Persistent requests for an addenda have been received based on the need to clarify the reasons for and/or the circumstances under which certain accidents or hazards occur and to include additional precautions and data. It was hoped that the article would generate wider recognition of the hazards, constructive comments to improve our hazard controls, and more case histories. Many cannot grasp the importance of safety precautions unless the reasons or consequences of unsafe procedures, processes and situations are illustrated by actual case histories or demonstration.

Statistically, the safe use of liquid and gaseous oxygen is apparently quite good. Its continued good record depends on improved identification and communication of high potential hazards. Although little information on liquid systems accidents is adequately communicated, a study published by *Anesthesiol-*

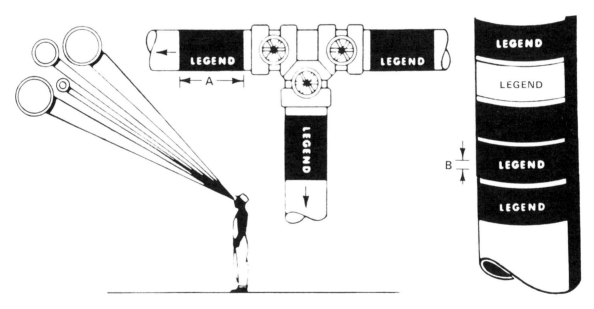

Legend Placement-Width of Color Bands and Size of Letters for Various Diameter Pipes

Method of identification
Positive identification of piping system content shall be by lettered legend giving the name of the content in full or abbreviated form. Arrows may be used to indicate the direction of flow. Where it is desirable or necessary to give supplementary information such as hazard or use of the piping system content, this may be done by additional legend or by color applied to the entire piping system or as colored bands. Legends may be placed on color bands.

Outside Diameter of Pipe or Covering	Width of Color Band A	Size of Legend Letters B
¾ to 1¼	8	½
1½ to 2	8	¾
2½ to 6	12	1¼
8 to 10	24	2½
Over 10	32	3½

ogy, January 1978, reports that one percent (.012) of compressed gas cylinders delivered to hospitals "had potentially hazardous irregularities in cylinder contents, identification, valves, testing, or storage."

Users must maintain close inspection surveillance and return every doubtful cylinder that does not have a clearly legible permanent label or stencil naming the contents, the proper valve outlet and inlet connections; or has any sign of leakage, or damage. Piping of liquid and gaseous oxygen must also be clearly labelled at the supply connections, every few feet along the pipe, and at discharge or output connections. (See American National Standards Institute ANSI Z48.1, "Marking Compressed Gas Cylinders" and ANSI B57.1, Compressed Gas Cylinder Valve Outlet and Inlet Connections".)

Sources of ignition

Sources of ignition of pressurized oxygen gas and liquid oxygen systems include sources beyond those of flame, electric arcs, and static discharge sparks. Other sources include compression wave fronts; pump, compressor, or other means of compression; friction, impact and other heat inputs.

Spills do not spontaneously ignite or explode. However, under accidental spill conditions, if there is any contaminant such as lubricant, asphalt, oil, ordinary grime, dirt, paint, etc. either on the surface under the spill or on a heavy object such as a hammer or wrench dropped on a LOX spill, explosion can occur upon impact. Another unique phenomenon is that explosions have occurred after LOX spilled into a crack in asphalt. Obviously asphalt should never be used where there is any possibility of LOX leaking on it.

It is a well established fact that both experience and adiabatic compression test apparatus shows that ignition energy can be added to a liquid by the heat of compression of pressure surges from quick-acting valves, the compression of liquid pumps, and even the compression of vapor bubbles. Such tests have not shown ignition energies from bubbles in liquid oxygen. Bubbles in liquids that are not confined, of course, cannot be compressed.

Van Dyke in the "Proceedings 1959 Cryogenic Safety Conference," reported that a series of mockup tests demonstrated the possibility of generating adiabatic compression wave front temperatures approximating the ignition temperature of stainless steel. He reported that was the explanation of several serious accidents in the missile industry which took place simultaneously upon the quick opening of shutoff valves against high pressure oxygen. Similar phenomena have occurred many times throughout general industry, when cylinder valves were opened upstream of regulators which were oil lubricated. The regulator metal bursts into flame. Friction or the impact of a particle of weld, a broken piece of a component or any foreign material may also serve as a source of ignition energy in a liquid oxygen or high pressure oxygen gas system. Open valves slowly. Consider small bypass lines for large valve systems.

Eliminate ignition sources

Avoiding any ignition energy source is vital since even stainless steel can burn in the presence of liquid oxygen when sufficient ignition energy is present.

Too many serious and even fatal burns have occurred in the many decades of oxygen use due to asbestos gloves, clothing, hospital bedding, etc. becoming saturated with oxygen. When any fuel and oxygen are mixed, sooner or later a source of ignition will cause clothing or any oxygen-saturated combustible to burst into flame and burn like dry straw or excelsior. Oxygen is not a safe substitute for air, which contains only 21 percent oxygen, because of the dramatic increases in speed of combustion once ignition takes place. Turning a switch on a lamp or a radio, static discharge from a person or materials, an arcing fan motor or switch, or a lighted cigarette—any of these can provide ignition. No ignition sources should be allowed in areas where oxygen is used or stored.

Consequently, while weatherproof types of electrical equipment may be the minimum that is required by codes, the extra insurance/hazard elimination of electrical equipment containing arcing devices is recommended by the writer. Also recommended are greater distances from fuel storage and from occupied buildings than those in the minimal requirements of the NFPA codes and OSHA Standards. Fire walls and other protective or preventive methods may be indicated.

Compressed gas cylinder thrust

The thrust available to move a compressed gas cylinder depends on the size of the gas passage hole that is opened when a valve is broken off. I have seen film of tests in which detonators were used intentionally to break valves. The cylinders barely moved. I could not determine if the detonators pinched the outlet or not. To obtain a velocity of 50 feet per second requires a one inch opening in a cylinder containing 2500 psi of oxygen. In many cylinders the opening left has been reported as being closer to a quarter of an inch which would explain the slow movement of some cylinders.

On the other hand, when "a welder's oxygen cylinder fell over and broke off the regulator (in a paper plant in 1974 in Menasha, Wisconsin), it was jet-propelled and struck and opened up a boiler temperating tank, releasing steam that fatally scalded seven men." In a conventional gaseous cascade oxygen storage system, holes burned in the cylinder ends, and several took off like missiles—landing 300 feet away. This occurred when the system was recharged to 2000 psig and the shutoff valve was closed. Experts in the industry said they were unable to determine the cause, although rapid adiabatic compression was a possibility. One authority concluded that adiabatic compression provided the energy of compression.

New code requirements

The ASME Code requirements have been refined and revised for various materials used at low temperatures. Since this code is also subject to very frequent interpretations by the ASME Code Committee, one must check with a code authority who keeps abreast of these changes to determine the current requirements.

Industrial practices

The design criteria published in the "Proceedings 1959 Cryogenics Safety Conference," for carbon steel lines in commercial oxygen service, a maximum pressure of 450 psig and a velocity restriction of 100 fps at 150 psig, was stated. It has been raised to 1000 psig at 25 fps and 250 fps at 150 psig as an industry standard in the Compressed Gas As-

WEIGHT AND VOLUME EQUIVALENTS

Liquid and Gaseous Oxygen and Nitrogen

OXYGEN

Weight of Liquid or Gas		Volume of Liquid at Normal Boiling Point				Volume of Gas at 70°F	
Pounds	Kilograms	Cubic Feet	Liters	Quarts	Gallons	Cubic Feet	Cubic Meters
1.000	0.454	0.0140	0.397	0.420	0.105	12.08	0.342
2.205	1.000	0.0309	0.876	0.926	0.231	26.62	0.754
71.27	32.327	1.0000	28.316	29.922	7.481	860.6	24.370
2.517	1.142	0.0353	1.000	1.057	0.264	30.39	0.861
2.382	1.080	0.0334	0.946	1.000	0.250	28.76	0.814
9.527	4.321	0.1337	3.785	4.000	1.000	115.05	3.258
8.281	3.756	0.1162	3.290	3.477	0.869	100.00	2.832
2.924	1.327	0.0410	1.162	1.228	0.307	35.31	1.000

NITROGEN

Weight of Liquid or Gas		Volume of Liquid at Normal Boiling Point				Volume of Gas at 70°F	
Pounds	Kilograms	Cubic Feet	Liters	Quarts	Gallons	Cubic Feet	Cubic Meters
1.000	0.454	0.0198	0.561	0.593	0.148	13.80	0.391
2.205	1.000	0.0437	1.237	1.307	0.327	30.43	0.862
50.46	22.888	1.0000	28.316	29.922	7.481	696.5	19.723
1.782	0.808	0.0353	1.000	1.057	0.264	24.60	0.697
1.686	0.765	0.0334	0.946	1.000	0.250	23.28	0.659
6.746	3.060	0.1337	3.785	4.000	1.000	93.11	2.637
7.245	3.286	0.1436	4.065	4.296	1.074	100.00	2.832
2.558	1.160	0.0507	1.436	1.517	0.379	35.31	1.000

sociation (CGA) Pamphlet G-4.4, "Industrial Practices for Gaseous Oxygen Transmission and Distribution Piping Systems." The reader should obtain it from the Compressed Gas Association, 500 Fifth Ave., N.Y., NY 10036 for more details.

The Compressed Gas Association has issued a Revised Pamphlet G-4.1 which is far superior to the original issue. It says that "harmful contamination would include both organic and inorganic materials such as oils, greases, paper, fiber, rags, wood pieces, solvents, weld slag dirt and sand which if not removed could cause combustion reaction in an oxygen atmosphere or result in an unacceptable product purity."

Carbontetrachloride is now prohibited, trichloreoethylene is no longer recommended and more definitive inspection details are given in the revised CGA pamphlet which serves as the industry voluntary standard. A note on the inside cover of G-4.1 cautions readers " . . . the Association . . . makes no guarantee of the results and assumes no liability or responsibility in connection with the information or suggestions herein contained. Moreover, it should not be assumed that every acceptable commodity grade, test or safety procedure or method, precaution, equipment or device is contained within, or that abnormal or unusual circumstances may not warrant or suggest further requirements or additional procedure." The same cautions and considerations apply to this author's articles.

Personnel precautions

Early experiences working with low temperature fluids indicated that direct and positive contact with most cryogenics could cause extremely rapid tissue damage somewhat similar to high temperature burns or frostbite. Researchers reported that prolonged exposure can embrittle and destroy exposed parts.

- Wear gloves which shed oxygen and which can be quickly removed, if necessary.
- Wear trousers with no cuffs so liquid cannot get into them.
- Wear chemical goggles or chemical face shields that have both top and bottom splash guards.
- Remove all clothing which has absorbed either liquid or gaseous oxygen and air it out for at least one half hour before it is considered safe to be worn again.
- No smoking, no flame, no ignition sources in oxygen handling areas.
- Never enter oxygen tanks for any purpose unless they have been purged and tested to ensure they have an atmosphere of breathing air of 19 to 21 percent oxygen
- Never wear greasy, oily or hydrocarbon soiled clothing or gloves in oxygen storage, handling or process areas.
- Avoid walking or stepping on any cryogenic spill.
- Use clean gloves or hands thoroughly soap-and-water-washed free of any oils, hand creams or soil when handling any oxygen equipment.
- Never hang hats, clothing or other objects on the valve end of a cylinder. It can become saturated from an unseen tiny slow leak.

Prevent overpressure

Ice cannot be permitted to form on relief valves or rupture discs. Valves must be properly calibrated, identified, inspected and maintained. Rupture discs should have permanently affixed tags and be designed to fail well before the container can fail.

Allison's findings on the hazards of using untagged and/or improperly rated rupture discs and the availability of threaded (non-flanged) disc fixtures for use with permanently affixed tags were published in 1960 in the *ASSE Journal* and abstracted in *Chemical Engineering News* in 1961.

At 70°F, liquid to gas expansion ratios are listed at 650 for methane, 780 for helium, 860 for argon, 875 for oxygen, 710 for nitrogen, 740 for air, and 865 for hydrogen by Zabetakis. The National Bureau of Standards Tech Note 361 lists 860 for oxygen, 695 for nitrogen and 848 for parahydrogen. (Some of there were erroneously shown on page 22 of January *PS*.) It should also be noted that silicone rubber is not recommended for use on oxygen systems and litharge and water is not a good sealant.

References

The following literature is available from the Compressed Gas Association, Inc., 500 Fifth Avenue, New York, NY 10036:

Handbook of Compressed Gases—Compressed Gas Association, Inc.
CGA Pamphlet G-1 Acetylene
CGA Pamphlet G-4 Oxygen
CGA Pamphlet G-4.1 Cleaning for Oxygen Service
CGA Pamphlet G-5 Hydrogen
CGA Pamphlet G-6 Carbon Dioxide
CGA Pamphlet G-8.1 Standard for Installation—Nitrous Oxide
CGA Pamphlet P-1 Safe Handling of Compressed Gases in Containers
CGA Pamphlet P-2 Characteristics & Safe Handling of Medical Gases
CGA Pamphlet V-1 Compressed Gas Cylinder Valve Connections
CGA Safety Bulletin SB-2 Oxygen Deficient Atmospheres

The following literature is available from the National Fire Protection Association, 470 Atlantic Ave., Boston, MA 02210:

NFPA No. 50 Bulk Oxygen Systems at Consumer Sites
NFPA No. 50A Gaseous Hydrogen Systems at Consumer Sites
NFPA No. 50B Liquified Hydrogen Systems at Consumer Sites
NFPA No. 53M Fire Hazards In Oxygen Enriched Atmospheres
NFPA No. 56F Nonflammable Medical Gas Systems
NPFA No. 69 Explosion Prevention Systems

"Better Things for Better Living Through Chemistry" was a familiar slogan a generation ago. Today's generation recognizes that some chemicals have side effects that interfere with better living. Two important laws are directed to preventing and controlling these side effects. These two laws are the Toxic Substances Control Act (PL94-469) and the Resource Conservation and Recovery Act of 1976 (PL94-580). These laws regulate the entire life cycle of a chemical: manufacture, distribution, use, and disposal. This paper presents examples of the events which led to this legislation, describes the legislation and current important issues, lists sources of information, and presents suggestions from the author's experience on how to comply.

Laws and regulations

Controlling toxic substances and hazardous materials

by George B. Stanton

Today we do have better things and do live better through chemistry and the products of the chemical industry. Synthetic fibers replace human tissue and create our easy-care wardrobes. Plastics are part of every phase of our lives—in transportation, communication, and industrial and consumer goods. Chemicals clean food-processing plants and hospitals, destroying or deterring harmful organisms and disease vectors. Chemicals increase the productivity of our farms and forests. The chemical industry makes a significant contribution to the national economy, with sales representing more than six percent of our Gross National Product. Millions of workers are employed by the chemical industry or chemical-dependent industries. Chemical sales now exceed $100 billion per year, with over 30,000 chemical substances in commerce. To these, a thousand new ones may be added each year.

Even though we do live better through chemistry today, chemicals and the industries that make and use them are under attack because some chemicals have negative long-range effects along with their immediate benefits. Here are some examples:

DDT—The louse killer of World War II was found to persist in the environment and to kill birds and fish—our first ecological crisis substance.

Kepone—An ant and roach pesticide causes nerve disorders and, perhaps, cancer in man. Fishing has been banned on the James River and parts of Delaware Bay because of kepone pollution.

Vinyl Chloride—This gas is the raw material for polyvinyl chloride (PVC), one of the most widely used plastics. Exposure to vinyl chloride now is stringently restricted by OSHA and EPA because of related liver cancers.

PCB's—Polychlorinated biphenyls were used as heat-transfer fluids in high-temperature processes, in some plastics, and as fire-resistant electrical insulation fluids in transformers, capacitors, and fluorescent light fixture ballasts. The PCB pollution in the Hudson River came from a plant making capacitors.

DBCP—Dibromochloropropane, one of the latest chemicals in the news, is a pesticide which kills tiny worms that infest vegetable fields. The same chemical now is believed to cause sterility in workers who process it.

These chemicals are called refractory. High temperatures e.g. 1,000°C (1,800°F) and long residence times (several seconds) are needed to decompose them in an incinerator. These same chemicals do not decompose in the environment, as do many other natural and synthetic organic chemicals. Instead, these refractory chemicals persist in the environment. The same molecules of pesticide that kill the insect or other pest then pass into the body of the predator who eats that insect or drinks the water in which the dead pest decays. The molecules of pesticide then accumulate in the predator and weaken or kill it, or pass on to its offspring, or pass to where they weaken or kill successive generations of aquatic life, after reaching a water course in the environment. These molecules of pesticide can be compared to lead pellets from the hunter's shotgun, that missed the water fowl rising from a pond. The lead pellets fall to the bottom of the pond and, in time, are swallowed by ducks, geese, or other bottom-feeders. The ingested lead then weakens or kills the feeder and its offspring.

Until the environmental movement took hold, we did not pay attention to the long-range consequences just described because we were not aware of these consequences. For many chemicals, we did not have the ability to detect and quantify micrograms (1×10^{-6} grams) or nanograms (1×10^{-9} grams)—and so measure concentrations of parts per million or parts per billion in the organs and body fluids of small animals or plants. Now that this analytical ability exists, scientists seek to relate the presence of a chemical to changes in organs, even in individual cells. These relationships, whether proven or speculative, have supported the environmental movement's shifting from requiring "clean air" and "pure water" to setting standards for the concentration of specific chemicals in the air we breathe, in the water we drink, and in the food we eat. The regulations on the use of fungicides, insecticides, and rodenticides (FIFRA) set by EPA and on food residues set by FDA are earlier examples of this shift. The Toxic Substances Control Act and the Resource Conservation and Recovery Act are the current results of this shift.

Toxic Substances Control Act

The Toxic Substances Control Act, (PL94-469), also known as TSCA or as TOSCA, was signed into law in October 1976, after many years of controversy in the Congress. Portions of this law were published in *Professional Safety*, December 1976. The purpose of this law is to anticipate and address chemical risks before it is too late to undo damage to human health and the environment. This law authorizes the Administrator of the U.S. EPA to require testing of chemicals (at the manufacturers expense), set pre-market notification of new chemicals, restrict (regulate) manufacture, use, etc., of chemicals, require record-keeping and reporting, require quality control procedures, and impose penalties. Inspections will be made. Disclosure of data is restricted; however, health and safety information is subject to disclosure. Any citizen may bring a civil suit under this law. The law applies to

TOSCA...
anticipating
and addressing
chemical
risks...

imports and exports, too. Chemicals used exclusively in pesticides, food, food additives, drugs, and cosmetics are exempted from the Act because all of the exclusions are regulated under other Federal laws. Chemicals used in cosmetics, exempt from TOSCA, will be reviewed by the Cosmetic Ingredient Review, an industry-supported program that expects to review between 10 and 50 ingredients, of the 2800 listed in the Cosmetics Ingredient Dictionary, each year. Priorities for review are based on frequency of use, use by children or the elderly, high concentrations, possible biological activity, and consumer complaints.

Initial chemical substances inventory

EPA's actions under TSCA have concentrated on the Initial Chemical Substance Inventory. This Inventory of existing chemicals is required under the law. A revised Inventory is to be published in 1980. The Inventory contains about 44,000 chemical substances in four volumes. Volume I contains the Initial Inventory arranged by Chemical Abstracts Service (CAS) Registry Number. Volumes II and III contain the Substance Name Section, an alphabetical listing of systematic chemical names and synonyms for substances on the Candidate List. Volume IV has two parts. The Formula Section orders substances with known chemical constitution by molecular formula. The Chemical Substances of Unknown or Variable Composition, Complex Reaction Products, and Biological Materials (UVCB) Section presents chemical substances that do not have specific molecular formula representations. Finally, there is a volume with Trademarks and Product Names and a volume with Reporting Companies. Supplement I was issued in October, 1979.

Manufacturers of chemical mixtures (e.g. paints) and research chemicals are exempt from these requirements unless the Administrator determines such reporting is necessary to enforce the Act. The same exemption applies to small manufacturers, except for chemicals that are subject to regulatory provisions of the Act.

Deciding who is a small manufacturer was a key element in EPA's proposed rules. Most safety and industrial hygiene professionals would agree that safety, health, and environmental problems occur more frequently among smaller manufacturers and the smaller plants in multi-plant corporations than in larger plants. In its first proposal, EPA would exempt companies that either manufacture at a single site and have annual sales of less than $100,000, or produce less than 2,000 lbs. a year of product. One industry group proposed another definition: annual sales of less than $30 million, less than 300 employees, or less than $415 million in assets. Adopting EPA's first proposed definition would have placed a great burden on companies short of scientific and managerial staff. On the other hand, adopting the industry definition would eliminate an unknown, large percent of the chemical industries' manufacturing capacity, and so, eliminate their chemicals from this regulation.

Small manufacturer, as defined in Section 710.2(x) of EPA's regulations on the Initial Chemical Substance Inventory, means a manufacturer whose total annual sales are less than $5 million based upon the manufacturer's latest complete fiscal year as of January 1, 1978. However, no manufacturer is a "small manufacturer" with respect to any chemical substance which such person manufactured in 1977 at one site in amounts equal to or greater than 100,000 pounds (45,400 kilograms).

Calculations for the $5 million criterion should be based upon the total sales of all products, whether or not they are chemical substances. In the case of a company which is owned or controlled by another company, the $5 million criterion applies to the total *annual sales of the owned or controlled company, the parent company, and all companies owned or controlled by the parent company taken together.* (Federal Register, *Dec. 23, 1977, pp. 64583,4)*

New chemicals—new uses

Manufacturers of new chemical substances must notify EPA at least 90 days before the manufacture of the chemicals for commercial purposes. Any chemical which is not listed on the initial inventory of existing chemicals will be considered "new." Premarket notification may also be required for significant new uses of existing chemicals. Premarket notification is now in effect. Limited exemptions are provided. The Act authorizes the EPA to prohibit or limit the manufacturing, processing, distribution, use, or disposal of a chemical pending acquisition of additional data. Similar requirements apply to drugs under FDA and such proceedings consume years before approval is given to market a new drug. The administrator may require fees, not more than $2,500, to defray the cost of reviewing testing data and premarket notifications.

Pre-market notification of 400 new chemicals each year was anticipated under this section of TOSCA. However, only 37 such notices were received during the last six months of 1979 and 42 in January 1980, for a total of 79.

Testing of chemicals

EPA may require manufacturers or processors of potentially harmful chemicals to conduct tests on the chemicals at their expense, to evaluate a chemical's health or ecological effects according to specified testing standards. The Inter-Agency Testing Committee, appointed from eight Federal agencies, has published lists of chemicals or mixtures for priority consideration for testing requirements. A group of similar chemicals may be one item on the list.

Industry is contesting such grouping of chemicals because each different chemical in a group can have different health or environmental effects from other chemicals in the same group. The health and environmental effects, which the tests are to identify and to quantify, go far beyond the classical LD_{50} (lethal dosage for 50% of the animals tested in a short-time test). These new tests seek subtle changes in the function of tissue and organs, the development of cancers, and other long-term exposure effects. Rather than doing retrospective epidemiological studies of workers exposed for a working life-time, tests over the life-times of two relatively short-lived species are conducted before exposing humans or the environment. Development of disease in animals is not directly transferable to humans. Thus, designing animal test protocols for a specific chemical and its metabolites, sometimes more toxic than the chemical itself, and then interpreting the results are tasks requiring experienced toxicologists, who are in extremely short supply. Such tests are expensive. For example, the cost of such tests for a pesticide is about $500,000. Laboratory testing capacity is limited; it will be placed under additional pressure if animal test data on pesticides, performed under earlier legislation, must be re-done because of recent allegations about the quality of that work.

These problems may be alleviated somewhat when EPA establishes their toxicological data system. Tests on single cells or other simple organisms, such as the currently popular Ames test, are another hope for the future.

EPA has proposed test rules and standards for test development for toxicity, chronic effects, and physical and chemical properties. See Figure 1, *Super-agencies regulate chemicals, following pages.*

Regulation (restriction) of chemicals

DUTY TO INFORM

"Any person who manufactures, processes, or distributes in commerce a chemical substance or mixture and who obtains information . . . that such substance presents a substantial risk of injury to health or the environment shall immediately inform the Administrator (of EPA)" (sec. 8 (e)). The Act provides protection from discrimination for employees who participate in carrying out the Act. The duty to inform applies to any person, including safety and industrial hygiene professionals.

EPA may prohibit or limit the manufacture, processing, distribution in commerce, use, or disposal of a chemical or mixture which presents an unreasonable risk to health or the environment. These restrictions may include process changes, quality standards, and quality control procedures. EPA may conduct inspections, too. Labeling may be required for a chemical or any article containing the chemical. When regulatory actions are proposed, there must be an opportunity for comments by interested parties, including an oral hearing, and in certain instances, cross-examination. For *imminent hazards,* EPA may ask a court to require action to protect against the risk.

Industry assistance office

EPA's Office of Toxic Substances has set up an Industry Assistance Office so those directly involved in or concerned about the manufacture, processing, distributing, use or disposal of chemical substances or mixtures, may be kept fully apprised of developments and have ample opportunity to participate in the developments. The Industry Assistance Office will also guide industry representatives to the proper EPA offices or executives. For more information and to be put on EPA's TOSCA mailing list, write:

John B. Ritch, Jr.
Director, Industry Assistance Office
Office of Toxic Substances
 (TS-557)
Environmental Protection Agency
401 M Street, S.W.
Washington, D.C. 20460
800-424-9065

Resources Conservation and Recovery Act

Millions of tons of waste are accumulating each year with less and less place for disposal. Landfill is becoming exhausted, pollution regulations restrict dumping on land and at sea, and burning pollutes the air. The most rational solution is to recover waste material through recycling. Part of the Resources Conservation and Recovery Act of 1976, passed almost as an afterthought to TOSCA, is directed to that end. But, hazardous waste management, another important part of the Resources Conservation and Recovery Act of 1976, is being enforced vigorously by EPA. EPA estimates 57 million tons of hazardous wastes are disposed of each year by 750,000 waste generators and that 90% of that waste is disposed of by environmentally unsound methods. Today, this is where the action is in the protection of public health from exposure to chemicals.

Hazardous waste management under RCRA

Hazardous waste management is the systematic control of all aspects

TOSCA Interagency Testing Committee

Section 4(e) of the Toxic Substances Control Act (TSCA) established the TSCA Interagency Testing Committee, to identify and recommend to EPA those chemical substances and mixtures to be tested to determine their hazards to human health and the environment. Factors to be considered by the committee when making recommendations include the availability of test data along with exposure and hazard potential.

This committee consists of representatives from:

Statutory Member Agencies

Council on Environmental Quality	National Institute of Environmental Health Sciences
Department of Commerce	
Environmental Protection Agency	National Institute for Occupational Safety and Health
National Science Foundation	National Cancer Institute
	Occupational Safety and Health Administration

Liaison Agencies

Department of Defense	Department of Interior
Food and Drug Administration	U.S. Consumer Product Safety Commission

From an initial list of about 3,650 chemical substances and categories, this committee made the following recommendations for testing.

Initial Report January, 1978	Second Report April, 1978
Alkyl Epoxides	Acrylamide
Alkyl Phthalates	Aryl Phosphates
Chlorinated Benzenes, (Mono- and Di-)	Chlorinated Naphthalenes
Chlorinated Paraffins	Dichloromethane
Chloromethane	Halogenated Alkyl Epoxides
Cresols	Polychlorinated Terphenyls
Hexachloro-1,3-butadiene	Pyridine
Nitrobenzene	1,1,1-Trichloroethane
Toluene	
Xylenes	

Figure 1. Super Agencies regulate chemicals

of a solid waste which may contribute to an increase in illnesses or deaths, or which may pose a hazard to health or the environment. A waste is hazardous if it is ignitable, corrosive, reactive, or toxic. Toxicity of a water extract (leachate), rather than the waste, is measured. Wastes already declared hazardous include pesticides and solvents. Processes generating hazardous wastes include textile dyeing and finishing and battery manufacture. Generators of hazardous wastes must comply with regulations on record keeping, labelling, containers, furnishing information to transporters, a manifest system, and reporting to the EPA or to a designated State agency. (New Jersey has a manifest system and is working with New York, Connecticut, Maryland, Massachusetts, New

-76-

Interagency Regulatory Liaison Group (IRLG)

In August 1977, the heads of the Consumer Product Safety Commission; the Environmental Protection Agency; the Food and Drug Administration; and the Occupational Safety and Health Administration formed the Interagency Regulatory Liaison Group (IRLG) to ensure that the agencies work closely together in areas of common interest and responsibility. One of eight groups established to promote better coordination was the Regulatory Development Work Group, which undertook to identify hazardous materials that two or more IRLG agencies planned to regulate; and ask the writers of these regulations to jointly prepare a development plan for regulating hazardous materials. Development plans for regulating 24 hazardous materials were published on December 1, 1978.

The 24 materials are:

Acrylonitrile	Ethylene Dibromide
Arsenic	Ethylene Oxide and its Residues
Asbestos	Lead
Benzene	Mercury and Mercury Compounds
Beryllium	Nitrosamines
Cadmium	Ozone
Chloroform and Chlorinated Solvents—Trichloroethylene (TCE); Perchloroethylene (PCE); Methylchloroform	PBBs
	PCBs
	Radiation
Chlorofluorocarbons (CFC)	Sulfur Dioxide
Chromates	Vinyl Chloride (VC); Polyvinyl Chloride (PVC)
Coke Oven Emissions	Disposal to Food Chain Land
Dibromochloropropane	
Diethylstilbestrol	

National Toxicology Program (NTP)

Established by the U.S. Department of Health, Education and Welfare in November, 1978, the NTP combines parts of four HEW agencies that study chemical's human health and environmental hazards. These four agencies are the National Cancer Institute, National Institute for Occupational Safety and Health, National Institute for Toxicological Research, National Institute of Environmental Health Sciences. These agencies committed $41 million from their Fiscal Year 1979 budgets to NTP. The level estimated for the 1980 fiscal year was $69 million. The NTP Executive Committee consists of heads of these four agencies and of the four IRLG agencies. Hundreds of chemicals are being tested for possible genetic toxicology, carcinogenisis, chemical disposition, general toxicology, immunologic toxicology, neurobehavioral toxicology, pulmonary toxicology, and reproductive and developmental toxicology.

Hampshire, Rhode Island and Vermont to interchange information.) Regulation of transporters is to be consistent with U.S. Department of Transportation Hazardous Materials Regulations. Regulations on treatment, storage, and disposal facilities will require assurances of financial responsibility and continuity of operation including a trust fund. Such facilities will be prohibited in high hazard areas e.g. flood plains and earthquake zones and will be required to have an EPA permit to operate. Inspections and taking of samples by the EPA are permitted.

This law provides for no discrimination against an employee who initiates or testifies in a proceeding under this law. Citizen suits are permitted, as is the awarding of costs of litigation.

The Clean Water Act and Amendments and regulations set by the EPA, Corps of Engineers and the U.S. Coast Guard severely restrict the concentrations of hazardous chemicals permitted in the waste water discharged to river or stream (receiving body) or to a waste water receiving (sewage treatment) plant.

Current enforcement

The field enforcement staff of EPA has identified thousands of active and abandoned hazardous waste disposal sites and is using the courts to impose fines and to compel safe disposal or containment. Litigation and Court appearances are handled by the U.S. Justice Department's hazardous waste section and by the local United States Attorney. These agencies believe hundreds of sites have the potential for major human health and environmental effects. The 1980 fiscal year budget for solid and hazardous waste management is being increased twelve million to about twenty-two million dollars. The 1981 fiscal year budget request is expected to be even higher. Other EPA programs have given up 235 staff members for transfer to EPA's hazardous waste program, both in Washington and in the regional offices. Enforcement is expected to be by State agencies, after adoption of EPA rules by that State.

In Niagara Falls, EPA sued Hooker Chemical Corp. for $124 million in connection with leachate migration from the highly-publicized inactive Love Canal and three other disposal sites.* In Edison, N.J., a partial settlement has been reached in the first case filed for violating the Resources Conservation and Recovery Act. Kin-Buc, Inc., agreed to deposit $500,000 in an escrow fund, to be used to pay for a cover over its 20-acre dump of hazardous waste, and to pay for air and water pollution monitoring. The cover will be a 2½-foot thick composite of plastic, clay, sand and earth to keep rain off the waste chemicals and prevent leaching into two aquifers.

Three Elizabeth, New Jersey, companies and nine individuals have been indicted for picking up 40 million gallons of untreated acid waste in New Jersey and nearby states, and dumping it through a hidden underground pipeline into

---

The State of New York is suing Hooker and its parent, Occidental Petroleum, for $635 million.

> **Toxic inferno strikes on Earth Day**
>
> Fire roared through the Chemical Control Corporation from 10:55 p.m. on April 21 through 9:00 a.m. on April 22, 1980, the Tenth Earth Day. The fire incinerated more than 1,000,000 gallons of hazardous waste stored at this Elizabeth, N.J., site across the narrow Arthur Kill from Staten Island. Hundreds of drums exploded.
>
> The heat of the fire and long residence time in the fire resulted in nearly complete combustion of more than half of the 35,000 drums of hazardous waste left after the controversial site was taken over by New Jersey's Department of Environmental Protection and the company's president was convicted of creating and maintaining a public nuisance. Favorable weather conditions and the billions of BTU's released by the fire sent the billowing smoke plume straight up into the night air for thousands of feet, so that the residents in the surrounding area had minimal exposure to the combustion products.
>
> Meanwhile, a bill to create a $600 million industry-financed "superfund" is moving through the Congress, accelerated by the Chemical Control Corp. fire.

Since the article on Controlling Toxic Substances was written by Mr. Stanton, the events above could impact the issue.

the Arthur Kill, a narrow channel separating New Jersey from Staten Island. One company and its president face fines of up to $1,557,000 and up to 157 years in jail.

RCRA rules

Record keeping rules for implementing RCRA were published in the *Federal Register* on February 26, 1980, to take effect in October. Additional rules, defining hazardous wastes and regulating their disposal, are scheduled for adoption in April 1980 (see above).

Record keeping rules require anyone who produces, transports, stores or disposes of hazardous wastes to apply for an EPA identification number. EPA is mailing out 350,000 application forms. The key to enforcement will be imposed on the producers, or generators, of hazardous wastes. They will be required to initiate a four copy manifest listing the specific facility to receive that waste. The manifest will identify the waste's source, nature, quantity, and the shipper. The manifest is nearly identical to the manifest required by the U.S. Department of Transportation for hazardous waste carriers. The manifest will go with the shipment, each recipient signing and keeping one copy, as the shipment is transferred from the generator, to the carrier, and, finally, to the EPA approved disposal site who would return the fourth copy to producer certifying to receipt of the waste. Follow-up after 35 days is required. In addition, waste producers must file annual reports with EPA.

Charges of improperly disposing of hazardous wastes, against eight companies and three of their officers, are still under litigation. The Justice Department is seeking $1.6 million in damages and penalties.

Midnight dumpers

Midnight dumpers are the bane of responsible chemical manufacturers and users and the modern hazardous waste disposal industry. Dumping was justified in the past by the dilution of the waste by flowing rivers, large lakes, and the vastness of the ocean. Dumping was cheap, a few cents a gallon; the used drum containing the hazardous waste was worth more than the dumping fee. Disposing of hazardous wastes in a manner that has no, or minimal, effect on health and the environment is expensive. Costs range from $1.00 to $10.00 per gallon. The new EPA manifest system is specifically directed against the midnight dumper.

Closing hazardous waste dumps, pressures to remove these wastes, community resistance to badly needed new hazardous waste disposal plants, and the continuing generation of new wastes creates more opportunities for "midnight dumpers." Trailers of hazardous wastes, abandoned overnight, are found on side streets and vacant lots in Newark and Elizabeth, New Jersey, and on unused farm land in Pennsylvania. Criminal charges, both felony and misdemeanor, have been brought by the State of Pennsylvania against eleven individuals for dumping as much as 2,400 gallons per day between August 1978, and July 1979, into an abandoned coal mine that drained into the Susquehanna River. The Susquehanna, a major river in East-

---

TEN STEPS TO CONTROLLING HAZARDOUS MATERIALS TOXIC SUBSTANCES

1. Read the labels on packages and containers you receive, warehouse, store, or use. One or more probably contains a substance hazardous to your workers, your plant, or the public.
2. Obtain safety data sheets from suppliers for materials and supplies which carry warning labels or bill of lading hazardous materials descriptions, and for chemicals, mixtures, synthetics, etc.
3. Make these warnings and instructions part of Job Safety Analysis and of the work procedures used by your workers.
4. Make a flow sheet showing how each hazardous material moves through your plant, including by-products and wastes. Confine these materials in quality equipment and facilities. Use closed systems for transfer and for processing; plan for waste disposal *before* wastes are generated. If you can't dispose of waste, the system will overflow, like a plugged sewer.
5. Make hazardous materials conservation part of your energy conservation program. Recover and recycle these materials back into your process. Recover fuel values in a wastes-fired boiler or pre-heater. Keep wastes of different types separate to ease analytical and disposal costs.
6. Find out how the USDOT hazardous materials regulations on packaging, labeling, and placarding apply to you as shipper of products, by-products, and wastes. Train your employees to comply.
7. If you sell, ship, import, or export chemicals, find out how the new Toxic Substances Control Act applies to you as an individual.
8. Prepare a spill prevention, control, and countermeasures (SPCC) plan to keep spills of toxic substances and hazardous materials from becoming pollutants and to preserve their dollar value.
9. Keep current through *Professional Safety,* trade magazines, newsletters, seminars, trade associations, and *Federal Register.*
10. Start today!

Figure 2.

"Dumping hazardous wastes on a plant's own property is prohibited unless... approved by EPA..."

ern Pennsylvania, is a source of water, fish, and recreation for perhaps one-quarter of the state. The cost of cleanup may be as high as $6 million. If convicted on all charges, each of the eleven individuals indicted face penalties of up to twelve years in prison and $50,000 in fines.

Keeping hazardous wastes separate will reduce in-plant costs of analyzing containers before disposal and the disposal costs. Dumping hazardous wastes on a plant's own property is prohibited unless done in a manner approved by EPA and permits will be required.

For current information, contact the USEPA's Assistant Administrator for Water and Waste Management (currently, Eckardt C. Beck), Washington, D.C. and your State's environmental protection agency. Books and seminars on hazardous wastes and their management abound. Because the field is changing swiftly, they, and this article, can go out of date as soon as the ink is dry.

Conclusion

Chemicals are an important part of our national economy. They are used, in one form or another, in every plant, office, and home. The Toxic Substances Control and the Resources Conservation and Recovery Act regulate the life cycle of a chemical. Figure 2 "Ten Steps to Controlling Hazardous Materials... Toxic Substances" will help safety professionals, engineers and plant managers comply with these new laws.

". . . packaged cargo can pose problems. . . . Nowadays, many shipments are lashed to pallets . . . thousands of materials are shipped in steel drums, glass carboys, glass jars, tin cans, paper bags, and all sorts of packages. . . . Weather can be a problem, too, since wet or icy roads greatly increase the chances of jackknifing or collision with another vehicle or a road hazard." Photo courtesy of American Chain & Cable Co., Bridgeport, Conn.

Transporting, loading, and unloading of hazardous materials using motor vehicles

by **William S. Wood**, CSP, P.E.,
Chemical Safety Consultant,
West Chester, Pennsylvania

ABSTRACT. Safety in motor truck transport of hazardous cargo requires both improved rolling stock and the upgrading of carrier personnel through training. Identification of hazardous materials, accurate shipping papers, proper loading and unloading procedures, and correct use of placards are concerns of DOT as well as the shippers and carriers.

Changes in regulations are being promulgated to simplify compliance and improve safety. Emergency information services such as MCA CHEMTREC have proved valuable in times of spill or fire.

Before talking about the transportation of hazardous cargo, it might be well to define what we mean by the term hazard. Generally, we think in terms of flammability, instability, toxicity, and corrosivity. Other hazards may be present, such as elevated pressures or temperatures, or very low temperatures (cryogenic substances). Some package cargo may be radioactive and it is possible that the shape and configuration of a cargo could render it hazardous. Consider, for example, a poorly packaged shipment of razor blades.

Most often the term hazardous cargo has reference to chemicals. And actually, if we carry it far enough, we would have to say that practically everything is chemical

*Reprinted with permission of The Society of Manufacturing Engineers. Paper presented at the Joint Materials Handling Conference held September 23-25, 1975, in Cleveland, Ohio.

"Operating about 10% of the rail mileage in the U.S., Southern Railway system has taken a most progressive action in the handling of emergencies. The company now has a "go team" of three men equipped with tools and protective gear transported to emergency sites in an automobile-drawn trailer. If an accident is more than 100 miles away, the trailer is flown to the nearest airport, from where it is pulled by an automobile to the scene of the emergency." Photo courtesy Southern Railway System.

in nature. Not long ago the American Chemical Society publication, *Chemical Abstracts,* classified and listed its three millionth compound. That means 3 million different chemical compounds have been isolated or synthesized and sufficiently identified for them to meet the rigid criteria of *Chemical Abstracts.* Although many of these are laboratory curiosities, this does not prevent samples being shipped by truck or by other means of transportation. Only a couple of hundred chemical materials, not all of them hazardous, are normally shipped by bulk carriers. But certainly several thousands of materials are shipped in steel drums, glass carboys, glass jars, tin cans, paper bags, and all sorts of packages.

Many materials have not even had their hazards determined. Toxicity or stability may be completely unknown quantities until the material has been in commercial use for some time. The number of materials listed by the Department of Transportation with specifications for their marketing and packaging is actually quite short when one considers the very large number of materials that may be encountered on the loading dock offered for shipment by common carriers.

Equipment

It is a basic consideration that transport equipment should be suitable to the particular material and the job it is called upon to perform. Let us just consider a tank truck that is to be used for a liquid material. DOT regulations specify the type of truck or trailer for particular cargos, but the specifications leave some discretion to the owner or carrier. The tank must be sufficiently strong to carry the weight and specific gravity of the material. You would not want to ship concentrated sulfuric acid in a tank truck designed for petroleum products, since the weight would be approximately twice what the truck has been designed for. The material of which the tank is constructed must be chosen to minimize corrosion and contamination.

The connections for filling and emptying the truck must be suited to the particular type of cargo connections encountered at the pickup and delivery points. The type of relief device must be suited to the particular cargo. A material that generates vapor when heated will need a different kind of device from the one that is not volatile. Another consideration is the protection of connections in case of a rollover or collision.

Even the simple van may be unsuitable for certain types of packages. For example, I know of cases where gas cylinders shipped in such a vehicle were not supported and were literally allowed to roll around in the van subject to the possibility of a cap being loosened and a valve broken off.

If the carrier is offered a mixed cargo, who is going to decide what materials are compatible? In case of an accident, packages may be breached; and resulting mixture of contents can threaten the whole van, if not the community.

Tank cleaning

Another problem, particularly in reference to tank vehicles, is cleaning between different types of cargo. Tank trucks haul a variety of materials, and the tank bottom must be adequately cleaned whenever the nature of the cargo changes in order to avoid contamination and also to avoid possible reaction between two different cargos. There is a documented case where a tank truck containing a substantial amount of glycerine had some nitric acid erroneously added to it. As you know, the result of this combination is nitro glycerine; and, believe me, the contents of that tank truck were removed very, very carefully. This cleaning operation is a very serious and expensive proposition, particularly since the washings may not be recoverable and can constitute a serious pollution problem. It is so expensive that a small individual operator can not afford to do as adequate a job of pollution abatement as he should. There seems to be some tendency for tank line operators in a given area to combine their cleaning operations in order to make the facility large enough to handle wastes properly.

Maintenance

Trucks are not always as well maintained as they should be. When they are in the shop they are not earning money, and the cost of repairs can be very substantial. For this reason, there is some tendency for smaller operators to keep their trucks running until a breakdown occurs or a number of mechanical items must be corrected. This procrastination can lead to breakdowns on the road with serious consequences. Consider what might develop if a cryogenic material had a limited amount of time for the transit and pressure was building up with the delay. Tires must also be maintained; it has been demonstrated many times that it is sheer folly to haul a hazardous cargo on a set of smooth tires.

Probably no part of the transport system is as important as the braking. Sudden stopping, particularly with wet roadway conditions, can cause jackknifing with dire consequences to the driver and the truck. Modern trucks are more and more being equipped with computerized braking so that the jackknifing potential is considerably reduced. Jackknife prevention linkages are available, but truck drivers do not like them because they allegedly reduce maneuverability of the rig. They are, however, effective and have saved lives when they were being used.

Inherently, both tank vehicles as well as vans have a high center of gravity, and this, of course, leads to the susceptibility to rollover when an unfavorable road condition exists or a sharp turn must be negotiated. Heavy package cargo should be loaded on the floor of the trailer in order to keep the center of gravity as low as possible.

Driver training and assignment

The suitability of the driver and his crew is even more important than the rig when a hazardous cargo is being carried. Paramount are the training and experience of the driver and his helper. A driver who has not had experience in carrying a hazardous cargo should not be made responsible for such a task until he has worked with a driver who has knowledge and experience with the particular cargo being carried. Helpers, of course, will be less experienced, but even they should have instruction by the operating firm before being sent on a job even with an experienced driver. I know of one case where an inexperienced driver and an equally inexperienced helper were sent to pick up and deliver 23 tons of concentrated sulfuric acid. In the process of making delivery by pressurizing the tank vehicle with air, the connection was prematurely broken and both men were sprayed with acid and suffered painful burns.

Much of the responsibility for assignment of the crew and also for warning them about material hazards and precautionary measures must lie with the dispatcher, and to do this properly he must have an awareness of the hazards of the cargo that is being handled. The dispatcher should see that protective clothing and other necessary equipment is available and is sent with the rig so that they can be used as necessary by the crew for their protection.

Gloves, eye protection, respiratory protective devices, etc.

It is not enough that protective equipment be available, but the crew must be trained in its use. They must know when and how to wear gloves, eye protection, raingear, and most of all, respiratory protective devices, gas masks, and the like. At least three and possibly more of the larger carriers have training programs for their drivers and other employees. One company has a trailer van equipped as a schoolroom, and outfits it with movies, pieces of typical equipment, and so forth. The trailer is then hauled to various locations and used to train the crews.

It is characteristic of the trucking industry that the driver is usually responsible for the unloading of his cargo. This is not true in rail, water, or air carriers. Standard operating procedures and experience cannot be overemphasized. There is one case on record where a tank truck driver obtained guidance as to the unloading point and thought that he was following the directions that were given him. Instead, he managed to discharge a

"Having taken on his cargo, picked up his shipping papers, and properly placarded his truck, the driver then sets out on that hazardous ribbon of concrete, the American highway." "He must negotiate curves, most of which are designed for passenger cars rather than trucks." Photo courtesy of Uniroyal.

"A recent piece of legislation that has had considerable impact... is the Federal Aid Highway Amendments of 1974. One of the major provisions of this legislation is the permanent establishment of the national 55-mile-an-hour speed law. While I know of no statistics that prove that the reduced speed limit enables us to transport hazardous materials with fewer accidents, it is generally conceded that the lower speed limit plus the reduced amount of driving results in a reduction in fatalities of 10,000 people annually."

cargo of nitric acid into a tank containing hydrochloric acid, and the resultant fumes killed several people in the plant as well as the driver. In another case, hydrogen sulfide asphyxiated several employees when acid was mistakenly delivered into a tank of sodium sulfide solution. Whoever was responsible, the driver was definitely involved. When a driver is picking up cargo, he must find the person who can tell him exactly what to take and can identify the contents for his information. Unfortunately, in the case of many chemical cargos, neither the driver nor the dock worker really has sufficient understanding of terminology to communicate adequately.

Shipping papers

It is the responsibility of the shipper to furnish completely filled out waybills and other shipping papers. Often the clerk who makes up these papers does not have the background to give the complete information expected, or even to identify the material properly. When the cargo does not fit the standard list of hazardous materials in the tariff, it may be simply designated "not otherwise identified," or some other catch-all terminology. The generic name, not the trade name, must be shown on the shipping papers, and it is proper and prudent to give any useful warning information. Knowledge that the material is toxic or corrosive may prevent someone from receiving an overdose or severe contact with it.

Another piece of information that should be on the shipping papers is the instruction to call CHEMTREC if there is any problem with respect to the cargo. I will say more about CHEMTREC later.

Placarding

If the cargo exceeds certain minimum weight limitations, a placard must be shown on the outside of the truck when hazardous materials are being carried The exact design and wording on the placard is spelled out by regulations of the Department of Transportation, and it is generally conceded that it is the responsibility of the shipper to make available the placard for attachment to the transportation vehicle. It is, however, the responsibility of the driver to verify that the placard matches the information on his shipping papers and affix the proper number of them in designated locations on his vehicle. It will be the carrier rather than the shipper who is fined if the incorrect placard is being displayed, or if the truck is not placarded when such is required.

Road hazards

Having taken on his cargo, picked up his shipping papers, and properly placarded his truck, the driver then sets out on that hazardous ribbon of concrete, the American highway. He must negotiate curves, most of which are designed for passenger cars rather than trucks. His speed on these curves often cannot be as great as those posted because of his high center of gravity. A high crown accompanied by an adverse curve has thrown many a truck over on its side. Weather can be a problem, too, since wet or icy roads greatly increase the chances of jackknifing or collision with another vehicle or a road hazard.

We have heard a great deal about the very poor maintenance of railroad beds, but all of you know that trucks are often required to negotiate very rough roads, very narrow roads, roads that are more suited to pack mules than to 40-foot multi-ton cargo tractor rigs.

Drivers are expected to get to their destination in a reasonable time. When delayed at the pickup point, they may attempt to make up lost time to avoid working overtime or being censured by their employer. Traffic, of course, varies greatly; and many a large truck has run into serious difficulty because a small passenger car, typically driven by a little old lady school teacher, pulled out in front of him.

Unloading

Well, let us say he does negotiate all of the curves and high crowns and wet roads and narrow roads, maintains a safe speed, and handles all the traffic problems, and then arrives at the destination where he is to unload. If there is a dock facility, this is fine for package cargo. Usually bulk deliveries are to pumps, piping systems, conveyer systems, or other arrangements for bulk handling. The driver must verify the unloading arrangement and secure approval for beginning the unloading operation. Sometimes this involves the setting of valves, or the connecting of air lines, and usually this must be done by the employees of the consignee.

Tank vehicles are unloaded both from the bottom and from the top. Some cargos are traditionally unloaded at the bottom while others are nearly always top unloaded. The unloading operation may be by pumping, using either a pump on the tractor or the customer's pump. However, a substantial number of unloading operations are performed by applying air or gas pressure to the top of the tank and actually pushing the contents out to the receiving tank by gas pressure. Maximum pressure may be specified by the Department of Transportation and should seldom exceed about 30 pounds per square inch. The time of unloading differs greatly, depending upon the pressure of the gas and the length and size of the discharge line. Unloading often takes an hour or more, and the driver is often impatient and anxious to speed up his unloading as much as possible. This can result in his taking shortcuts or applying excessive pressure which can be quite hazardous.

The driver must know when to wear protective clothing and must wear it whenever necessary. Failure to do this, despite the discomfort that it may cause in hot weather, can result in his suffering severe injury from contact with the cargo. A safety shower and water supply must be present when a hazardous material is being handled. This, of course, is the responsibility of the consignee, but it is the job of the driver to know where the shower is and to be sure he knows how to use it before he starts the transfer operation. In the sulfuric acid splash I mentioned a while ago the driver ran past a shower that was only 20 feet from his truck and went at least a hundred feet to a much less adequate water supply to be washed off. The driver must be sure that his vehicle is empty and depressurized before he disconnects his lines, even though this takes a little bit longer.

The unloading of packaged cargo can pose problems. Nowadays, many shipments are lashed to pallets and are loaded and unloaded by forklift. It is not too unusual for a fork truck to damage packages and to spill the contents, resulting in potential contact with the material.

Emergencies

When a serious accident occurs during transit, there is the possibility of spill, release, reaction, fire, or explosion of the cargo. The mere threat of these consequences is enough to motivate local police and fire services into prompt action. Rarely do they have training or experience upon which to base their strategy. Hopefully, they will have access to the shipping papers; but if the tractor cab is involved, they may have no information other than the warning on the placard. Some emergency officers carry with them copies of the *NFPA Guide on Hazardous Materials* or the *MCA Chemcard Manual*.

For the past four years an information service called CHEMTREC has been available to help in transportation emergencies. In 1971 the Manufacturing Chemists Association set up, in Washington, D.C., a 24-hour Chemical Transportation Emergency Center. CHEMTREC has collected information on the hazards and emergency handling of all products shipped by the 167-member companies of the MCA. A fire chief or other officer can call CHEMTREC on a toll-free number — 800/424-9300 — and give the product name from the shipping papers. Within 10 to 20 seconds he can expect to obtain the properties of the material and simplified directions for emergency action. If the shipping papers are destroyed or missing, the truck number and carrier may be useable by CHEMTREC to trace the cargo back to its shipper and obtain vital information. CHEMTREC is not a government-operated agency but a voluntary service furnished by a Trade Association and its dedicated membership. In the last four years it has responded thousands of times to calls for help from all over the world.

A unique and useful pamphlet for the emergency services was published in 1974 by the Office of Hazardous Materials of DOT, titled, "Emergency Services Guide for Selected Hazardous Materials." Based on approximate spill area and threshold limit value, the booklet recommends evacuation patterns for releases of 19 common toxic materials. Similarly, it gives ignition control distances from spills of 11 commonly shipped flammables. Immediate action information is given for each material.

Regulation

No discussion of highway transportation would be complete without some consideration to the regulatory agencies that affect these operations. Although there are a number of state and local regulations, we will, because of their universal effect, confine our remarks to U.S. government agencies and their regulatory activities. The Department of Transportation has historically been responsible for certain phases of the shipment of hazardous cargo, at least since the time when DOT absorbed these functions from the Interstate Commerce Commission.

Fairly recently, Kenneth Pierson, Deputy Director of the Bureau of Motor Safety, Federal Highway Administration, in a speech before the American Society of Safety Engineers, said, "With the advent of the Occupational Safety and Health Administration, the Consumer Product Safety Commission, the National Fire Prevention and Control Administration, and other new regulatory agencies joining the host of traditional or historical safety, health, fire, police agencies, there is probably no area of human endeavor now left unprotected in this country."

The Transportation Safety Act of 1974 (Public Law 93-633) substantially increases the jurisdiction and scope of government regulation of transportation of hazardous materials. Title I, the Hazardous Materials Transportation Act, is intended to improve the regulatory and enforcement authority of the Secretary of Transportation to protect the nation from risks of life and property inherent in the transportation of hazardous materials in commerce. Some of the more important aspects of Tile I are:

(1) Centralizing under the Secretary of Transportation the authority to administer regulations governing the packaging, labeling, handling, storage, and transportation of hazardous materials by any mode. This includes, but is not limited to, explosives, radioactive materials, etiological agents, flammable liquids and solids, combustibles, oxidizing materials, corrosives, poisons, and compressed gases.

(2) The authority of DOT is no longer limited to interstate commerce but also covers transportation within the state.

(3) Extending coverage by regulation to all shippers of hazardous materials and also to manufacturers of containers intended to be used in complying with DOT regulations.

(4) The Act establishes standards for the qualifications

and training of persons handling hazardous materials and also specifies the procedures and facilities necessary to assure protection of persons coming in contact with such hazardous materials. (It should be pointed out here that the Department of Transportation maintains in Oklahoma City a training school where any transportation employee may be sent to learn the principles of handling hazardous cargo.)

Since the Act was signed into law by the President on January 3, 1975, implementation of the Act has hardly begun, and its effect so far is barely perceptible.

Title III of the Act is a very interesting one. The purpose of this Title is to separate the National Transportation Safety Board from the Department of Transportation. Up until the Act was passed, the National Transportation Safety Board, while a part of the Department of Transportation, actually reported directly to Congress rather than to the Secretary. Hence, it had a great amount of independence; and this was intended when it was created by Congress. The purpose of the National Transportation Safety Board is to investigate accidents in an impartial and thorough manner and to gain as much information as possible. The lessons learned can be applied to the future avoidance of such accidents or attenuation of their effects once they do occur. The Act now gives NTSB an unequivocal freedom to carry out its investigations and make its reports on a completely unbiased basis.

Another recent piece of legislation that has had considerable impact, and will have more, is the Federal Aid Highway Amendments of 1974. One of the major provisions of this legislation is the permanent establishment of the national 55-mile-an-hour speed law. While I know of no statistics that prove that the reduced speed limit enables us to transport hazardous materials with fewer accidents, it is generally conceded that the lower speed limit, plus the reduced amount of driving, results in a reduction in fatalities of 10,000 people annually.

A recent regulation that will have major effect upon safety in transportation of hazardous materials is the National Highway Traffic Safety Administration's new brake standard for trucks equipped with air brakes. MVS 121 should result in better braking ability and improved stability of large commercial vehicles, since it reduces braking distances without deviation from the travelled lane.

Some of these regulations cost money, and there is some thought that they should be modified or postponed in view of the economy and the possible effect upon inflation. On this topic, William T. Coleman, when he was being considered for confirmation by the Senate Commerce Committee, stated that we should not expect serious consideration of a rollback in the Federal Highway Program as long as we are experiencing more than 40,000 highway fatalities each year.

Another rulemaking action that is presently under consideration is the consolidation of hazardous materials regulations that were formerly issued under three separate titles of the Code of Federal Regulations. Obviously, the problems of coordinating terminology for transport by rail, highway, waterway, and air is a very ambitious undertaking. Those of you who are familiar with Graziano's Tariff 27 may not believe that such can be done. If so, I would suggest that you obtain from the Department of Transportation a copy of Docket HM-112 which appeared in the Federal Register, Thursday, January 24, 1974, Volume 39, No. 17, Part II.

Summary

In closing, I would like to quote a statement developed by the Safety and Fire Protection Committee of the Manufacturing Chemists Association. It is: "Chemicals in any form may be safely stored, handled, or used, if the physical, chemical, and hazardous properties are fully understood and the necessary precautions, including the use of proper safeguards and personal protective equipment are observed."

This is a true statement, and yet we know that there are gaps in our knowledge, there are those who do not have the necessary information, there are those who will not take the proper safeguards. Regulations are of little effect unless they are obeyed, and compliance can best be achieved by training people in the meaning of the regulations and the steps needed to comply with them. New and advanced equipment is of value only if it is properly used, and its proper use must come about through training. Hazards of materials are not so complex that they can not be readily understood, but training is required so that people may understand the hazards and the protective measures that are needed to prevent injury and disaster. –PS–

about the author

William S. Wood is a Chemical Safety Consultant in West Chester, Pennsylvania. He joined Sun Oil Company as a Research Engineer after receiving a bachelor of science degree in chemical engineering from Purdue University. He was a Safety Engineer at Sun Oil for 20 years before setting up his consulting business. He is a Certified Safety Professional and Professional Engineer.

Mr. Wood has served on committees for the Manufacturing Chemists Association and the American Institute of Chemical Engineers. He has co-edited a book on safety. In 1973 he organized a curriculum in occupational safety and health at Northeastern Christian Junior College, and teaches in the program. He was elected a Society Regional Vice President, Region XII, in 1972 and was re-elected in 1974. Mr. Wood was a speaker at the Society's 1975 Professional Development Conference held in Denver.

Professional Safety:

Ventilation Index:
An easy way to decide about hazardous liquids

A Ventilation Index based on the TLV is derived and is shown to be more useful than the TLV alone. This Index is of great value to the safety professional in making prompt and sound decisions about the safe use of liquids and the required ventilation criteria when these liquids are used in various industrial or chemical processes. Moreover the use of this Index is easy and does not require sophisticated instrumentation.

by Azmi P. Imad, and Cheri L. Watson

Safety professionals and industrial hygienists are often faced with situations like the following:

A facility, chemical process or experiment, uses several organic liquids in a certain environment whereby the evaporation of these liquids cannot be prevented. The safety professional is called upon to evaluate the hazard involved and to prescribe measures that assure safe working conditions.

Principles of control

In such a case, three of the basic principles of control are available to the safety professional. These are:

First: Try to find substitute liquids which are safer than those used by the process, or

Second: Determine which of the liquids in use present the worse hazard and recommend their deletion, or methods to control their vapors, or

Third: Determine ventilation rates which can dilute the vapors of the most hazardous liquids or any other liquids, to safe working limits.

While focusing on finding measures of control according to the above three principles, the readers should be informed that these principles are not the only methods of dealing with such a problem. Other methods such as the elimination of the process, its change, isolation in time or space, using personal protective equipment etc., are also possible. It is also important to note that fire, explosion or other hazards need consideration along with the health hazards which are considered here.

For the sake of simplicity, and in order to focus on the value of the Ventilation Index, ideal conditions such as steady states, uniform vapor emission, perfect mixing etc., are assumed in the following examples and discussions.

The need for a new approach

Industrial hygiene books that are commonly available do not usually provide adequate treatment of this kind of problem, and when they do, this treatment will often involve the use of some form of detecting equipment. Other literature and information that the safety professional may have, such as TLV/TWA tables[1] are not adequate if they are to be used separately from other references.

The Ventilation Index that we shall describe here provides the safety professional with a powerful tool that enables him to pass sound judgment in relation to the three above mentioned basic principles of control. Moreover using this index is simple and requires the use of little or no tools. On many occasions this index has proven to be more valuable than the TLV/TWA tables.

Conventional method and ventilation index approach

In order to illustrate how this index

can be used, let us first give more details to the situation described earlier and second, see how we can make decisions on the application of each of the three principles of control. In doing so, we shall compare the use of the Ventilation Index method to those methods that are conventionally being used.

Assume the facility in question has several columns and containers which are filled with liquids (see Figures 1 and 2). These containers cannot be

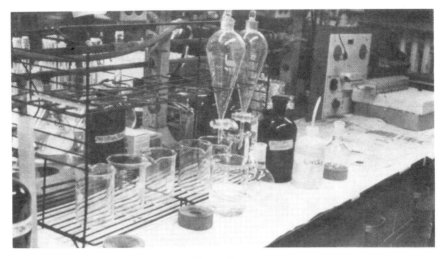

Figure 1
Chemical process setups such as these, using columns or containers that cannot be closed, do not prevent the evaporation of liquids.

Figure 2
Chemical process setups such as these, using columns or containers that cannot be closed, do not prevent the evaporation of liquids.

closed, hence vapors will be emitted into the atmosphere. Let us also assume that one of the following four liquids has to be used in this process:

toluene
dipropylene glycol
ethyl benzene
ethyl formate

From a health hazard point of view, and whether we are using conventional methods or the Ventilation Index, we have to determine the amount of vapors that each liquid would put into the atmosphere if we are to arrive any where near an answer to any of our earlier questions. The following conventional methods can be used in solving this problem.

Conventional Methods of Control

A. *Direct Measurements:*

Several methods involving the use of simple and sophisticated equipment exist for measuring toxic vapor concentrations in air (see Figure 3).

However, whatever method is used the measurements should extend over a large span of time ranging from a few days to several weeks, in order to make sure that the obtained values of vapor concentrations in the atmosphere are true representative values that are not affected by short temporary fluctuations in environmental conditions and/or changes in the experimental process.

B. *Indirect estimates*

Another way to attack this problem is to find out from the user what amounts of liquid are being lost as vapors. This can be obtained from the records that the user may have such as amounts purchased, amounts put into the process, amounts disposed of and amounts remaining after the process is finished. However, it is often the case that the user is unable to give true representative values in such instances. If the information is provided, the user usually admits that these values are difficult to measure and record and that they are only rough estimations (see Figure 4, next page).

The direct measurements as in A will yield results which can be readily converted into parts per million (ppm). From this, one can easily evalute the degree of hazard with the help of TLV/TWA tables[1] or other toxicology references.

The indirect estimations as in B above usually result in an amount of liquid (such as pounds, pints, gallons, etc.) that is evaporated in a certain time. This amount can be easily converted into a ppm concentration using the equation:

$$\text{ppm} = \text{lbs liquid evap} \times \frac{392}{\text{MW}} \times \frac{10^6}{V} \quad (1)$$

Where MW is the molecular weight of the liquid, and V is the volume in cubic feet of the room or enclosure into which vapors are released. Equation (1) assumes a temperature of 75°F and that there is perfect mixing of the air, and that no ventilation is occurring.

Knowing the vapor concentration in ppm, we can easily determine the amount of air needed to dilute these

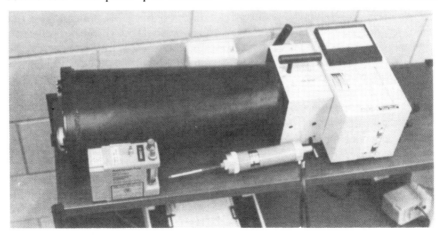

Figure 3
Infrared gas spectrophotometer, sampling pump for use with charcoal tubes, and gas detector tubes, are some of the equipment used to determine toxic vapors in the air.

concentrations to any desired level. This level can be the Ventilation Design Concentration (VDC)[2], the TLV or any other safe level. The following equations give the amounts of air needed in cubic feet per minute (cfm), that are needed to dilute vapor concentrations to the VDC and TLV respectively:

cfm required to achieve VDC =

$$\frac{\text{lbs liquid evap}}{\text{Time in minutes}} \times \frac{392}{MW} \times \frac{10^6}{VDC} \quad (2)$$

Where VDC is expressed in ppm, and

cfm required to achieve TLV =

$$\frac{\text{lbs liquid evap}}{\text{Time in minutes}} \times \frac{392}{MW} \times \frac{10}{TLV} \quad (3)$$

The cfm obtained from the above equations should be multiplied by a safety factor[3] which depends on the toxicity of the liquid, the method of ventilation chosen, the conditions and location of the process.

Use of the conventional methods A and B allows us to come up with ventilation rates needed for the use of each liquid. These ventilation rates can be used to indicate which liquid is the most hazardous to use. Hence, if we apply methods A and B to each of the four liquids mentioned, we can answer the three basic questions posed earlier. However to perform measurements on four or several liquids would be extremely time consuming. This would be particularly more difficult if the number of liquids is increased to say 10 or even 20. In such cases we can use the Ventilation Index method. This method is a lot easier to use, does not require the use of sophisticated equipment and consumes much less time.

Ventilation Index method

Noting that

lbs liquid evaporated × 392
　　　　　　　MW

= cu. ft. (or vapor volume)

equation (3) can be rewritten as follows:

cfm required to achieve TLV =

$$\frac{\text{cu. ft.}}{\text{Time in minutes}} \times \frac{10^6}{TLV} \quad (4)$$

Figure 4
It is often difficult for the user to determine amounts of liquid evaporated in a process such as the one shown.

or

$$\text{cfm required to achieve TLV} = \frac{Vv}{E} \times \frac{10^6}{TLV} \quad (5)$$

Where Vv is the volume in cubic feet of the vapor produced by evaporating the liquid at 75°F, and E is the evaporation rate which is the ratio of the time required to evaporate a certain volume of one liquid to the time needed to evaporate the same volume of a reference liquid under identical conditions. (For the purpose of comparing the hazards of various liquids we shall arbitrarily consider the volume of liquid in question to be one gallon. Hence Vv will be defined, in future equations, and tables as the volume of cubic feet of vapor produced by evaporation of one gallon of the liquid at 75°F.) Using ethyl ether as the reference liquid and using a reference volume of ethyl ether which needs one unit of time to evaporate (say 1 minute), E can be expressed as the time needed to evaporate a certain volume of any liquid substance. Equation (5) uses E in this context. In this case E is referred to as the evaporation time (in minutes) relative to ethyl ether for which E = 1.

Equation (5) shows that the vapor health hazards of a liquid depend on three variables which are Vv, E and the TLV. Obtaining ventilation rates will always involve the multiplication of the factor

$$\frac{Vv}{E \times TLV}$$

by a constant number. Hence it is this factor which we shall call the Ventilation Index which determines the degree of hazard that the use of each liquid presents. This Ventilation Index (V.I.) is therefore defined by the equation

$$V.I. = \frac{Vv}{E \times TLV} \quad (6)$$

Going back to our hypothetical situation, and using the TLV-TWA tables, one finds the following values:

	TLV-TWA
toluene	100
dipropylene glycol	100
ethyl benzene	100
ethyl formate	100

These values of TLV's are identical and might mislead the novice safety professional in assuming that these four materials are equally hazardous when used in the liquid state. Calculating the Ventilation Index for these liquids reveals the following:

Compound	TLV	Vv	E	V.I.
toluene	100	31	4.5	.0689
d. glycol	100	25	300	.0008
ethyl-benzene	100	27	9.4	.0287
ethyl formate	100	41	1.8	.2278

Using equation 5, we find that the volume of air needed to dilute the vapors from each liquid to the TLV would be the product of V.I. × 10⁶ cu. ft.

For the four liquids, these values are:

Compound	Cu. ft. of air required to dilute the TLV
toluene	68,900
dipropylene glycol	800
ethyl benzene	28,700
ethyl formate	227,800

Using the Ventilation Index, we can now answer the 3 questions that were posed earlier.

1. If it was found that any of the 4 liquids was suitable for the process, then the safety professional can easily recommend the use of dipropylene glycol as a substitute for the other 3 liquids. This is so because dipropylene glycol has the lowest Ventilation Index and therefore requires the least amount of air for ventilation purposes.

2. The answer to the second question would certainly indicate that ethyl formate is the most hazardous liquid

to use in this process. The safety professional might decide against the use of ethyl formate and perhaps toluene the second most hazardous substance. These decisions will of course depend on the volume of the enclosure and the amount of ventilation available.

3. In order to control the vapors of the most hazardous substance, which is ethyl formate, we would need 227,800 cubic feet of air to be mixed with the vapor produced from each gallon of ethyl formate. This ventilation rate will reduce the concentration of ethyl formate to the TLV of 100 ppm. Moreover, this ventilation rate is also more than adequate to control the vapors of the other three liquids when used in the same process under the same conditions.

Useful applications of Ventilation Index

In the above example, the four materials were chosen because they had equal TLV's and therefore would simplify the use of the Ventilation Index. However, one can generalize the use of this index to any material for which the three factors, TLV, Vv and E are available[4], for example, if the safety professional was to consider a process which could use one of the liquids mentioned below, then it would be easy to make recommendations based on the value of the Ventilation Index using the following table:

Figure 5
Simple methods such as the one shown above, can be used to determine evaporation rates in the same environment as that of the process.

concentrations should also be taken into account. In order to evaluate the dilution ventilation rate needed for use with each of the above mentioned substances, one can use simple methods to determine the rate of evaporation of any liquid in the same environment of the process in question. (See Figure 5.)

This liquid could be a safe liquid for which the values TLV, Vv, and E are known. Once the actual evaporation rate under the condition of that particular experiment are known, then

Compound	TLV	Vv	E	V.I.
tetrahydrofuran	200	40	2	.1000
acetic acid	10	57	11	.5182
aniline	5	36	200	.0360
benzene	10	37	2.8	1.3214
benzenyl chloride	1	28	48	.5858
carbon disulfide	20	54	1.6	1.6875
ethanolamine	3	33	>5000	<.0022

The safest liquid to use is ethanolamine. Considering the TLV's alone would seem to indicate that tetrahydrofuran is the safest liquid, since it has the highest TLV and is safest, while ethanolamine has the second lowest TLV and is therefore the second most toxic substance.

In considering ventilation requirements or in choosing which is the safest substance to use, other factors, such as odor, flammability, containment of liquid, and ventilation design

the evaporation rate for any other liquid can be determined from the evaporation tables[4]. Using the Ventilation Index, one can thus determine the ventilation rate that would be needed for the use of any liquid in the process.

The industrial hygienist is instructed that the use of any liquid whose vapor TLV is less than 100 ppm, requires local exhaust ventilation rather than dilution ventilation. According to this, the reader should be informed that for most of the liquids mentioned above, one might recommend local exhaust ventilation. On the other hand, we would like to bring to the attention of safety professionals that tetrahydrofuran, whose TLV is 200 ppm, has a Ventilation Index much higher than those of aniline (TLV = 5 ppm) and ethanolamine (TLV = 3 ppm). Therefore it is also wise to use local exhaust ventilation for tetrahydrofuran since its use in the liquid state is more likely to result in a vapor concentration above the TLV, than the use of the other two liquids.

Table of ventilation indices

Following is a tabulation of ventilation indices for some 90 compounds that might be used in several ways by safety professionals who may not have access to sophisticated measuring instruments. This table was found to be of great value by the authors in situations where a sound and prompt judgment on the choice of the compounds and their required ventilation rates were needed.

Note: The ventilation index is a good method to determine relative hazards. It will not totally eliminate the need to perform measurements. Local ventilation can have stagnant areas. Some vapors are heavier or lighter than air and tend to concentrate in layers, and unless ventilation is properly designed, concentration may be different for the different vapors. The index should be used with caution. Also the TLV isn't always a measure of hazard. Some TLV are set on such things as odor, irritation etc., in addition to toxicity.

References

1. American Conference of Governmental Industrial Hygienists, *Threshold Limit Values* for Chemical Substances in the Working Environment with Intended Changes for 1977.
2. Olishifski, Julian B. and McElroy, Frank E., General Ventilation and Special Operations, Fundamentals of Industrial Hygiene, Chicago, National Safety Council, 1971.
3. Sax, Irving N., Industrial Air Contaminant Control, Dangerous Properties of Industrial Materials, New York, Van Nostrand Reinhold Co., 1975, pages 62-63.
4. Handbook of Organic Industrial Solvents, American Mutual Insurance Alliance, Chicago, 1972, 4th Edition.
5. Based on data from reference 1, 4, and Table of Chemical Hazards, Accident Prevention Manual for Industrial Operations, National Safety Council, Chicago, 1969.

TABLE OF HAZARD INDICES[5]

Substance	TLV (ppm)	Vapor Volume cu ft per gal	Evaporation Rate (Ether = 1)	Ventilation Index
acetaldehyde	100	58	3.0	0.1933
acetic acid (glacial)	10	57	11.0	0.5182
acetic anhydride	5	35	17.4	0.4023
acetone	1000	44	1.9	0.0232
acetonitrile	40	62	5.2	0.2981
allyl alcohol	2	48	8.5	2.8235
allyl chloride	1	40	2.0	20.0000
n-amyl acetate	100	22	11.6	0.0190
iso-amyl alcohol	100	30	38.1	0.0079
aniline	5	36	200	0.0360
benzene	10	37	2.8	1.3214
benzyl chloride	1	28	47.8	0.5858
1-butanol	100	36	19.6	0.0184
n-butyl acetate	150	25	7.8	0.0214
n-butylamine	5	33	5.1	1.2941
carbon disulfide	20	54	1.6	1.6875
carbon tetrachloride	10	34	2.6	1.3077
chlorobenzene	75	32	8.2	0.0520
chloroform	25	41	2.2	0.7455
1 chloro-1 nitropropane	20	32	17.9	0.0894
o-cresol	5	32	>400	<0.0160
p-cresol	5	32	>500	<0.0128
cumene	50	23	13.9	0.0331
cyclohexane	300	30	2.6	0.0385
cyclohexanol	50	31	150 (approx)	0.0041
cyclohexanone	50	32	22.2	0.0288
diacetone alcohol	50	26	60 (approx)	0.0087
1,2 dibromoethane	20	38	12.4	0.1532
o-dichlorobenzene	50	29	39.7	0.0146
1,2 dichloroethane	50	42	3.3	0.2545
1,2 dichloroethylene	200	43	2.5	0.0860
dichloroethyl ether	5	28	200 (approx)	0.0280
1,1 dichloro-1-nitroethane	10	32	12.7	0.2520
1,2 dichloropropane	75	33	3.7	0.1189
diethylamine	25	32	2.2	0.5818
o-diethyl phthalate	5 mg/m³	17	·300	0.1124
diisobutyl ketone	25	19	30.8	0.0247
dimethylaniline	5	26	90 (approx)	0.0578
dimethyl phthalate	5 mg/m³	20	·200	0.1734
dioxane (diethylene dioxide)	50	39	5.8	0.1345
dipropylene glycol	100	25	·300	0.0008
epichlorohydrin	5	42	17.0	0.4941
ethanolamine	3	33	·5000	0.0022
ethyl acetate	400	33	2.7	0.0306
ethyl alcohol	1000	56	7.0 (approx)	0.0080
ethylbenzene	100	27	9.4	0.0287
ethyl ether	400	31	1.0	0.0775
ethyl formate	100	41	1.8	0.2278
ethylene chlorohydrin	5	49	23.9	0.4100
ethylenediamine	10	40	·5000	0.0008
ethylene dichloride	50	42	3.3	0.2545
ethylene glycol monomethyl ether	0.2	41	21.1	9.7156
furfural	5	39	60.0	0.1300
n-heptane	500	22	2.7	0.0163
n-hexane	500	25	1.9	0.0263
isophorone	5	22	200 (approx)	0.0220
mesityl oxide	25	28	8.4	0.1333
methyl acetate	200	41	2.2	0.932
methylal	1000	37	1.5	0.0247
methyl alcohol	200	80	5.2	0.0769
methyl butyl ketone	100	27	8.1	0.0333
methyl chloroform	350	32	2.7 (approx)	0.0339
methylcyclohexane	500	26	3.1	0.0168
o-methylcyclohexanol	50	27	140 (approx)	0.0039
methylcyclohexanone	50	27	38 (approx)	0.0142
methylene chloride	200	51	1.8	0.1417
methyl ethyl ketone	200	36	2.7	0.0667
methyl formate	100	53	1.6	0.3313
methyl propyl ketone	200	31	4.0	0.0388
morpholine	20	9	36.9	0.0122
naphtha (coal tar)	100	37	29-35 (ap-prox)	0.0106-0.0128
nitrobenzene	1	32	160 (approx)	0.2000
nitroethane	100	46	7.5	0.0613
nitromethane	100	61	6.6	0.0924
1-nitropropane	25	37	10.9	0.1358
2-nitropropane	25	36	7.7	0.1870
octane	300	20	5.9	0.0113
n-pentane	600	29	0.97	0.0498
n-propyl acetate	200	28	4.8	0.0292
iso-propyl acetate	250	28	3	0.0373
n-propyl alcohol	200	44	7.8	0.282
iso-propyl alcohol	400	43	7.7	0.0140
propylene dichloride	75	33	3.5	0.1257
pyridine	5	41	8.2	1.0
stoddard solvent	100	20	4.4 (approx)	0.0455
styrene monomer	100	28	12.4	0.0226
1,1,2,2 tetrachloroethane	5	31	19.1	0.3246
tetrachloroethylene	100	31	6.6	0.0470
tetrahydrofuran	200	40	2.0	0.1000
toluene	100	31	4.5	0.0689
trichloroethylene	100	36	3.1	0.1161
turpentine	100	18	375 (approx)	0.0005
vinyl acetate	10	35	3.3	1.0606
o-xylene	100	27	9.2	0.0293
m-xylene	100	27	9.2	0.0293
p-xylene	100	27	9.9	0.0273

Pesticide safety and health

C. Saran, Linda L. Sims, Paul M. Hughes, and Stephen A. Adam

According to the National Clearinghouse for poison control 355 individuals were poisoned by pesticides in 1978. This accounted for 4.8% of all the poisonings for the year. Pesticides ranked 6th in the total number of poisonings. During the same year there were 352 calls made to the Children's Mercy Hospital in Kansas City, Missouri, concerning pesticide poisoning. Within the last 10 years there were 59 hospitalized cases of pesticide poisoning accounting for 3% of all hospitalized poisoning cases.

Eight major categories of pesticide poisoning are:
1. Pesticide combinations
2. Insecticides
3. Rodenticides
4. Fungicides
5. Herbicides
6. Moth balls
7. Animal repellants
8. Insect repellants

Major modes of pesticide intake are:
1. Accidental ingestion
2. Kicks or trips
3. Inhalation
4. Other—skin, eyes
5. Self poisoning—unknown intent, gesture, and suicide
6. Unknown

Of these, inhalation and other are the most common modes of intake in agricultural operations. These modes of intake may not be reported as poisoning but result in cumulative and synergistic effects of the farm population. Most of the reported poisoning cases involve children under five years of age. For all ages pesticides account for 24.8% of inhalation and 14% of other poisonings.

Various safety and health legislations regulate the manufacture, distribution, sale, and use of pesticides. These legislations, enforced and monitored by various local, state, and federal agencies, intend to protect the farmers, farm workers, and other users from the hazards of pesticides. Environment Protection Agency (EPA) has extensive certification programs for commercial and private pesticide applicators. Re-entry times for various pesticides are specified. Federal Insecticide, Fungicide, and Rodenticide Act (FIFRA) requires that the workers be warned not to enter a field without protective clothing. National Institute for Occupational Safety and Health (NIOSH) recommends the use of personal protective equipment for pesticides and other hazardous materials.

Rationale
Central Missouri State University at Warrensburg, Missouri, is located in a farming community in the midwest. Pesticide safety and health takes personal dimensions in the community. Availability of personal protective equipment to the users of pesticides is essential in fulfilling the intent of pesticide safety and health legislations. In view of the grim statistics and the personal nature of the problem it was decided to investigate the availability of personal protective equipment.

Classes of pesticides
Pesticides are legally classed as "economic poisons" in most state and federal legislations, and are defined as "any substance used for controlling, preventing, destroying, repelling, or mitigating any pest." Pesticides can be classified according to their composition as follows:

Inorganic pesticides: These pesticides are most often made from minerals such as arsenic, boron, copper, lead, mercury, sulfur, tin, and zinc. Examples: lead arsenate, Bordeaux mixture, and Paris green.

Synthetic organic pesticides: These man-made pesticides contain carbon, hydrogen, and one or more other elements such as chlorine, phosphorous, and nitrogen. Examples: 2,4-D, atrazine, captan, parathion, and malathion.

Living micro-organisms: These include viruses, bacteria, and fungi produced by man. Examples: the bacterium Bacillus thuringiensis, and the polyhedrosis virus.

Plant-derived organic pesticides: These are made from plants or plant parts. Examples: rotenone, red squill, pyrethrims, strychnine, and nicotine.

Formulation of pesticides

Formulation of the pesticides is the final physical condition in which the pesticide is sold for use. Formulation improves the storage, handling, application, effectiveness, or safety characteristics of the pesticides. The term formulation is usually reserved for commercial preparation prior to actual use and does not include the final dilution in the application equipment.

Some common formulations, representing most of the pesticides used in the United States, are:

- Sprays (insecticides, herbicides, fungicides)
- Dusts (insecticides, fungicides)
- Aerosols (insecticides, disinfectants, or "germicides")
- Granulars (insecticides, herbicides, algicides)
- Fumigants (insecticides, nematicides)
- Impregnates (insecticides, fungicides, herbicides)
- Fertilizer combinations with herbicides, insecticides, or fungicides
- Baits (insecticides, molluscicides, rodenticides)
- Slow release insecticides
- Insect repellants
- Insect attractants
- Animal systemics (insecticide, parasiticides)
- Animal dressings (insecticide)

Toxicity and hazards of pesticides

The toxicity of a pesticide is not synonymous with its quality of being a health hazard. Toxicity is the capacity to produce injury or harm while hazard is the possibility that the exposure will cause injury when a specific quantity is used under certain conditions. Some key considerations in determining the health hazard of pesticides are:

1. How much of the pesticide is required to be in contact with a body cell and for how long to produce injury?
2. What is the probability that the pesticide will be absorbed or come in contact with body cells?
3. What is the rate of generation and purging of the airborne contaminants?
4. What protective devices or other control measures are in use?

The effect of exposure to pesticides depends on the dose, rate, and physical state; site and duration of absorption; and diet and general health of the individual.

In developing a pesticide for experimentation and exploration, toxicity data are collected on pure pesticides. The tests are conducted on animals that are easy to work with and whose physiology, in some instance, is like that of humans; for example, white mice, white rats, white rabbits, guinea pigs, and beagle dogs.

Intravenous tests are usually performed on mice and rats whereas dermal tests are conducted on shaved rabbits and guinea pigs. Acute toxicity determinations are most commonly made on rats and dogs, with the test substance being introduced directly into the stomach by a tube. Chronic studies are conducted on the same two species for extended periods, and the compound is usually incorporated in the animal's daily ration. Inhalation studies may involve any of the test animals but rats, guinea pigs, and rabbits are most commonly used.

All of these procedures are necessary to determine the overall toxicity. This is defined by LD_{50}, expressed as milligrams (mg) of toxicant per kilogram (kg) of body weight, or the dose that kills 50 percent of test animals. Eventually, some microlevel portion of the pesticide may be permitted in food for humans as a residue, which is expressed in ppm.

Pesticide toxicologists use rather simple animal toxicity tests to rank pesticides according to their toxicity.

Before registering a pesticide with the Environmental Protection Agency and the eventual release to the general public, the manufacturer must declare the toxicity of the pesticide to the white rat under laboratory conditions.

Effects on humans

Most commonly reported symptoms of pesticide contact, in their order of progression, are:
  Headache
  Visual disturbance—blurred vision
  Pupillary abnormalities—primarily pinpoint pupils but on rare occasions dilated pupils
  Greatly increased secretions as:
    Sweating
    Salivation
    Respiratory secretions
More severe poisonings result in:
  Nausea and vomiting
  Pulmonary edema—fluid in the air spaces of the lungs
  Changes in the heart rate
  Muscle weakness
  Respiratory paralysis
  Mental confusion
  Convulsions
  Coma
  Death

Even if the prognosis for "recovery" is good in the cases of prompt treatment, there may be some permanent damage to the nervous system resulting in a drop in the I.Q. after hypoxic seizures.

The pesticide label

With the application for registration with EPA, the pesticide manufacturers are required to submit data which shows that the product, when used as directed, will be effective against the pests listed on the label and will not cause unreasonable adverse effects to man, animals, crops, or the environment. This informa-

> EPA... requires that certain statements appear on all product labels.

tion is utilized to construct label statements that warn pesticide users of potential hazards and give precise directions on how to apply the pesticide. EPA also requires that certain statements appear on all product labels. It reviews all labels for accuracy before approving the pesticide for public sale. It is against the Federal law to use any pesticide in a manner inconsistent with the label directions.

The label should contain the following key elements, in addition to the product's name, ingredients, and the quantity:

1. EPA registration number
2. Directions for use
3. Precautions—including hazards to humans and domestic animals, and environmental and physical or chemical hazards. The most toxic products will be labeled DANGER-POISON. The word WARNING on the label means that the product is less toxic but extreme care must be exercised in its use. The word CAUTION will appear on those products which are least harmful when used as directed. Particular attention should be paid to the warnings about keeping the children and pets out of treated areas and about special clothing and personal protective equipment.
4. First aid instructions—for accidental poisoning
5. Storage and disposal
6. Classification statement—including re-entry and warranty statements. "Restricted" use pesticides are highly toxic or require special knowledge or equipment for application and, therefore, should not be available to the general public. "General" use pesticides are available to the general public and should be used as directed.

Data collection

This study investigated the awareness and the incidence of pesticide safety and health problems among the major users of the pesticides—the farmers and the availability of proper personal protective equipment.

The questionnaire (Figure 1) was developed and administered to a random sample of 100 farmers in Johnson County, Missouri. The farmers were contacted by telephone by the students of Central Missouri State University. Various pesticide dealers and farm supply stores in Warrensburg, Odessa, and Sedalia were visited to select a pesticide and obtain the needed personal protective equipment.

Results

The results of the survey of the farmers (Table 1) indicated that:

1. Ninety-four percent of the respondents stated that they always read the instructions and the precautions on the label.
2. Sixty-eight percent of the respondents always followed the instructions, 22 percent followed them most of the time, and 10 percent followed them sometimes.
3. Seventy percent had not received any formal instruction in the safe handling and application of pesticides.
4. Eighteen percent of the respondents, members of their families, or others who had worked for them had become ill after using or handling pesticides.
5. Ninety-two percent were aware of the health hazards from the improper use or handling of pesticides.
6. The most objectionable aspects of using safety equipment were:

    Safety equipment interferes with work activity 22%

Figure 1—QUESTIONNAIRE

1. Age: _____ years.
2. How long have you been farming? _____ years.
3. Do you use pesticides? YES: _____ NO: _____
4. Do you read the instructions and the precautions on the pesticide label? YES (always): _____ NO (never): _____ SOMETIMES: _____
5. Do you follow the instructions concerning application and safety precautions? YES (always): _____ NO (never): _____ SOMETIMES: _____ MOST TIMES: _____
6. Please explain why you do or do not follow the instructions and safety precautions. _____
7. Have you received any formal instruction in handling an application of pesticides? YES: _____ NO: _____
   If YES, where and by whom? _____
8. Have you, members of your family, or any person who has worked for you ever become ill after using or handling pesticides? YES: _____ NO: _____
9. Are you aware of the health hazards associated with the improper use or handling of pesticides? YES: _____ NO: _____
10. Which of the following most accurately describes your objections to the use of safety equipment?
    Safety equipment interferes with work activity: _____
    No safety equipment (gloves, respirators, etc.) on hand: _____
    Cost of safety equipment: _____
    Takes too much time: _____
    Other reasons: _____
11. If you had all the required safety equipment on hand, would you use it and follow safe procedures? YES: _____ NO: _____ MAYBE: _____
12. Who would you contact for help or information concerning pesticides? _____

Table 1—SURVEY RESULTS

1. Average age:	33 years
2. Average length of farming:	13 years
3. Do you use pesticides?	YES 83%
	NO 17%
4. Read the instructions and precautions on the label:	YES (always) 94%
	NO (never) 0%
	SOMETIMES 6%
5. Follow the instructions concerning application and safety precautions:	YES (always) 68%
	NO (never) 0%
	SOMETIMES 10%
	MOST TIMES 22%
6. Reasons for not following the instructions and safety precautions:	
Already familiar with the product. Too much time. Don't worry about it. Company's responsibility. The more the better. Bothersome. Expense. Habits. Don't know how, and Equipment may be more harmful than good.	
Reasons for following the instructions and safety precautions:	
Best results for crops. Safety. Cost of over-application. Health, and Contamination.	
7. Formal instructions in safe handling and application of pesticides:	YES 30%
	NO 70%
8. Farmers, members of their families, or others working for them becoming ill after using or handling pesticides:	YES 18%
	NO 70%
9. Awareness of health hazards from improper use or handling of pesticides:	YES 92%
	NO 8%
10. Most objectionable features in the use of safety equipment:	
Safety equipment interferes with work activity	22%
No safety equipment on hand (gloves, respirators, etc.)	44%
Cost of safety equipment	18%
Takes too much time	16%
Other reasons	0%
11. Inclination to use safety equipment and follow safe procedures if all the required safety equipment were on hand:	YES 52%
	NO 2%
	MAYBE 46%
12. Sources of help: Dealer, Extension Office, School	

No safety equipment (gloves, respirators etc.) on hand 44%
Cost of safety equipment 18%
Takes too much time 16%

7. If all the required safety equipment was on hand, 52 percent would definitely use it and 46 percent may use it.

8. Most commonly cited reasons for *not* following the instructions were:
   Equipment may be more harmful than good
   Already familiar with the product
   Company's responsibility
   Don't worry about it
   The more the better
   Don't know how
   Too much time
   Bothersome
   Expenses
   Habit

Most commonly cited reasons for *following* the instructions were:
   Safety
   Health
   Contamination
   Cost of over-application
   Best results for the crops

The results of the visits to the farm supply and pesticide stores indicated that:

9. In 50 percent of the stores that sold pesticides no protective equipment was available.

10. The remaining 50 percent of the stores did not have adequate protective equipment.

11. One store had one cartridge respirator in stock and numerous filters but did not carry gloves which would be adequate.

12. Another store had a number of replacement filters but no respirator in stock.

13. While there was a large variety of work gloves available none was adequate for handling pesticides.

14. Splash goggles were not available and the investigators were told that welders' goggles were.

Analysis of results

From the results of this study it is evident that a large majority (92 percent) of the respondents were aware of the health hazards associated with the improper use or handling of pesticides, yet only 68 percent said that they always used safety precautions and read the instructions for application carefully. One reason for this attitude may be the fact that only 30 percent had received any formal instruction in the safe handling and application of pesticides. Other reasons as stated by the farmers are listed in result 6 which are natural human responses.

The fact that 62 percent of the respondents felt that safety equipment costs more than what it was worth or had no safety equipment on hand to apply pesticide indicates:

1. Lack of education in the safe handling of pesticides and the difficulty in understanding the long-term cumulative and synergistic effects of pesticides, other agricultural chemicals and dust, and biohazards. The reason for this lack of understanding is that most of the pesticides do not show any symptoms immediately after short term exposures.

2. Unavailability of adequate personal protective equipment. This was further verified by visits to the area's farm supply stores.

Fifty-two percent of the respondents stated that they would definitely use safety equipment and 46 percent stated that they might use it if it was readily available and if they had the equipment on hand. Among the users of pesticides it was found that there is a definite need for readily available personal protective equipment.

If a farmer was interested initially in obtaining adequate protective equipment, he would have lost his interest after running all over town and not being able to obtain it. If he was sold protective equipment, there would be doubt as to its adequacy since it would have to be patched together from what was available here and there.

Personal protective equipment

In an effort to alleviate the problems

encountered by the farmers and applicators, readily available low-cost personal protective equipment kits were developed. They provide protection for head, eyes, face, breathing, and the body in one package thus eliminating the searching, hunting, and improvising usually associated with obtaining these items individually. The basic difference between the two kits is the disposability of the TYVEK coverall.

Kit contents and cost:
- Reuseable kit: Vinyl Suit ($3.50), Respirator ($16.65), Gloves ($1.41), and Splash Goggles ($2.20). Total Cost: $23.76
- Disposable kit: Tyvek Coverall ($2.67), Disposable Respirator ($2.67), Disposable Gloves ($0.49), Splash Goggles ($2.20). Total Cost: $8.03

Content Description:
- Vinyl Jacket: Snap-on collar permits attachment of hood. Hood, complete with drawstring, furnishes maximum protection. Roomy sleeves with take-up snaps on wrist, concealed snap closures down front of the jacket, ventilated cap, and air-vented under arm. Heat-sealed.
- Vinyl Pants: Drawstring waist pants, take-up snaps on cuffs for added protection. Heat-sealed.
- DuPont's Tyvek: Disposable coveralls made of spunbonded polyethylene, resistant to acids, bases, and salts. Complies with flammability standards. Coverall is sized for wear over conventional garments.
- Gloves: Nitrile gloves outperform natural rubber and neoprene in aromatic petroleum, and chlorinated solvents, and offer excellent resistance to gas diffusion and water permeation. They resist abrasion, cuts, snags, and punctures. Satinized, easy on and easy off. Silicon free.
- Goggles: Lightweight pliable plastic frame fits most facial contours comfortably with indirect ventilation which inhibits particle and mist infiltration.
- Respirator: NIOSH/MSHA approved for organic vapors, pesticides, paints, lacquers, and enamel mists, and dust and mists.

Specific contents of the kits may change but the concept of disposable and reuseable kits would be a significant contribution in providing for and gaining acceptability of readily available personal protective equipment. Specific kits for specific pesticides may also be developed. It is expected that the kit, when marketed on a large scale, will be sold for well within $15.00. Some kits would be much cheaper depending upon the contents and the use of pesticides.

Conclusion

It is hoped that kits for specific pesticides will be available for sale by the sellers of the pesticides. Another approach would be to distribute a general purpose kit as developed in this study and then sell specific items for specific pesticides.

Marketing of such kits will overcome the dilemma now faced by the farmers when attempting to purchase protective equipment adequate for the job. The pesticide users will be provided a complement of protective equipment that is adequate, readily available, and reasonable in cost. ✥

Presented at the American Industrial Hygiene Conference, Portland, Oregon, May 29, 1981.

References

Kansas State University. Pesticide safety depends on you. Cooperative Extension Service, pp 13-15, 1974.

National Safety Council. Fundamentals of industrial hygiene. Chicago, Illinois pp 8-9, 1979.

University of Hawaii. What you should know about pesticides. Cooperative Extension Service, Leaflet 139, 1972.

U.S. Agriculture Extension Service. Insecticides. Extension Bulletin 430, 1978.

—Tests and problems for use in private pesticide application. Extension Bulletin, 430-431.

U.S. Department of Agriculture. Apply pesticides correctly. USEPA Manual, 1976.

—Safe use of agricultural and household pesticides. Handbook Number 321.

U.S. Environmental Protection Agency. Farmer's pesticide use decisions and attitudes on alternate crop protection methods. 1974.

—Pesticides—read the label first. EPA 335, 1977.

—What you should know about the pesticide law. Handbook 243, June 1973.

Ware, George. The pesticide book. W. H. Freeman and Co., San Francisco, pp 1-137, 1978.

Wayland, J. Hayes. Safe use of pesticides. EPA Leaflet No. 322, USDA, 1977.

Health hazards in the workplace

By Ernest Mastromatteo

There is a distinct difference between toxicity and hazards. Toxicity is an inherent property of a substance, such as its boiling point or its vapor pressure. It's something that can be measured. The commonest method of measuring a substance's inherent toxicity, in order to determine its ability to produce harmful effects, is to test the chemical on experimental animals. A hazard is the likelihood that a substance will cause harm under specific conditions of use. This means taking into account the substance's physical properties as well as its inherent toxicity. The word safety is often used in this context. To me, safety means the complete absence of risk. I think we have to dwell on the term safety and what it really means, particularly when some people ask that the work be completely safe.

A chemical material has a typical dose-response curve. As a person is exposed to increasing amounts of a substance, effects are produced. Effects may be chemical in nature only and completely reversible with no adverse effects. There are also physiological responses without adverse effects. Up to a point on the curve, the effects are always completely reversible without any harm to the individual. But when we exceed that point we get irreversible harm and ultimately death.

In a typical dose-response curve, the normal homeostasis mechanisms in the body are able to compensate for this without any adverse effects. But, at a certain point, the effects become irreversible and may cause serious illness or death. So, when we talk about the toxicity of an agent, we have to first have an agent which is capable of inducing effects and a biological system which will respond. A toxic response is one that is deleterious to the system. I think it is important to understand this principle when testing possible cancer-causing agents on animals.

If we want to study environmental cancer in animals, we have to give them enough of the agent to produce a response. If we don't give enough to produce a response, we cannot talk of a response to cancer-causing substances. There is a very big argument about the value of giving high doses of these agents to animals. Can cancer induced in animals by administering high doses be applied directly to our understanding of cancer in man? There are two schools of thought on this question. One school says it is not meaningful because the experimenter involves other metabolic pathways in the animal which are not relevant to man. The other school maintains that giving large doses is the only way we can test whether these substances are really cancer causing. Otherwise, we would have to use millions of animals at low doses.

Toxic effects

A range of toxic effects is possible. They can be acute, representing the response to a single, large dose of a chemical. To illustrate an acute response, imagine drinking a bottle of whiskey at one sitting. You would certainly get an acute response within 24 hours. A sudden massive exposure to chlorine or sulphur dioxide will also result in acute effects. Chronic toxicity refers to the effects of small, repeated exposures to a chemical. One ounce of whiskey a day for 40 years will not produce acute effects but it may produce chronic effects. The same thing is true of day-to-day exposure to a chemical in the workplace.

The effects of a toxic substance may be local, that is they may be restricted to the area contacted by the agent, resulting in skin reactions or irritation of the mucous membrane. Contact with the teeth can cause erosion of the dental enamel as has been known to happen with acid mists. Systemic effects result when the agent is picked up by the blood and carried to organs like the liver, kidney and brain or to the nerve tissue. The effects on the organ or nerve tissue are remote from the place of inhalation or skin contact.

Chemicals encountered in the workplace can also be classified as irritants, allergens, carcinogens, mutagens, teratogens, etc. Of course carcinogenic and mutagenic effects are tied together because most people believe that in order to induce cancer changes in living cells, you have to exert a mutagenic effect, i.e., alter the genes of the cells. The relationship between mutagenesis and carcinogenesis explains why people are excited about the Ames Test which detects mutagenic changes in bacteria. Teratogens are agents which damage the developing embryo or fetus in the mother's womb.

Two fast-developing fields are genetic engineering and recombinant DNA. Generally, recombinant health hazards in the workplace are classified by either the effect produced or by the agent. I feel we need a combination of the two, so I usually classify them by agents and effects. There are biological agents. Take the example of a job that requires a person to work in a meat packing plant where the employee is exposed to viruses in the bodies of the animals. The person might get a viral disease which is truly occupational.

Route of entry

How are workers exposed to hazardous materials and how do they get into the body? The two conventional routes of entry into the body that are most important in occupational exposure are inhalation and skin contact. However, there are less common ways. Lunchrooms that are close to work areas could result in employees eating hazardous materials. Exposure by injection is also possible but rare. A nurse may prick an employee's finger with a syringe that is contaminated with virus or a mechanic working on diesel injection engines may inject diesel fuel into his subcutaneous tissues.

The first conventional route of entry, inhalation, is important. To be inhaled, the chemical or substance must be suspended in the air. These are called particulates and they can be in a solid or liquid form. They are easily inhaled and are a common cause of occupational disease. They also exist over a wide range of substances.

The types of solid particulates an employee may be exposed to are dusts, usually from frictional sources, e.g., grinding wheels; fumes, e.g., from molten metal which has been allowed to condense from the vapor; and fibres. The area of fibres is one place where the hygienist fights with the minerologist because the hygienist defines a fibre as anything that is at least three times as long as

The two conventional routes of entry into the body ... are inhalation and skin contact.

it is wide. This could include all kinds of things such as a typical asbestos fibre or fibrous glass, but it could also include long, needle-like objects which might be found suspended in the air. The types of liquid particulates an employee may be exposed to are fog, mist and droplets.

Particulates of the order of 20 microns are deposited in the anterior chamber of the nose. They stream by the turbinate bones of the nose and particles about 10 microns are deposited behind the turbinate in the nose and pharynx (the cavity leading from the mouth and nasal passages to the larynx and esophagus). The particles travel down the trachea at a velocity of 150 centimetres per second. Particles of about five microns drop out in the trachea and bronchial tubes. The velocity diminishes as the particles pass down the bronchi and the bronchial divisions, until we finally get to the small alveolar sac in the lung. At the individual air sac, where the exchange takes place, the velocity is close to zero and particles in the range of half a micron to three microns settle out in the alveoli.

Entrained particles

These are important aerodynamic functions since the respiratory system is designed to bring the air velocity in the lungs to practically zero so the gas exchange can take place. Entrained particles of half a micron to three microns that are suspended in the air will drop out depending on these aerodynamic conditions. This helps to explain why large particles drop out in the nose. An employee working with iron dust or carbon dust who blows his or her nose after work will see red or black dust and may wonder what is getting into his or her lungs. Particles that drop out in the nose can't get into the lungs because they are too large. However, a lot of other particles are capable of getting down because of their aerodynamic properties.

It's also important to understand how these particles are cleared. In the nose, we get rid of them in one to 12 hours by sneezing, blowing or picking.

The tracheo-bronchial system gets rid of dust in about 24 hours. In the bronchial system, there are columnar cells lining the tracheo-bronchial tube. They are arranged like columns with hair-like cells on them, interspersed every so often with a goblet cell which secretes mucus. There is a blanket of mucus to trap the particles and the waving cilia or hairs elevate these particles to the back of the throat. Once there, the person swallows the particles and they either enter the digestive tract or get spit out. This mechanism is a wonderful way of clearing dust. Unfortunately, cigarette smoking, as well as some occupational exposures to dust, damage it. If people smoke, the columnar cells tend to flatten, the mucus gets thicker and the hair cells tend to disappear so that the mechanism can't do the job it was designed to do.

When we get to the lung itself, there are a number of ways that particles can be removed. Soluble dusts are readily removed by the blood. Insoluble dusts are engulfed and removed by special cells called phagocytes or macrophages. However, some dusts set up an inflammatory reaction which is later expressed as scar tissue, e.g., silica and asbestos. The scarring usually takes many years to develop. Scarring by asbestos is different from scarring with silicosis. An occupational health expert trained in reading X-rays can distinguish between these two types of scarring.

In some cases, we even have situations where there is exposure to mixed dust. For example, employees may be exposed to both silica and asbestos when making asbestos-cement pipes. With this kind of exposure, you may have the chest X-ray showing the classical evidence of mixed-dust pneumoconiosis. The mineral talc has many common uses but many things are called talc. Some forms of talc may have asbestos fibres in them. Therefore, it is very important for a person to know what they are talking about when referring to some of these minerals.

Gases and vapors

There is also a wide range of gases and vapors which can be inhaled and possibly cause adverse health effects. Some cannot be seen, have no odor and are non-irritating so employees are not aware of their presence. Carbon monoxide is a good example. It is found wherever there is combustion, e.g., gasoline engines, propane engines and fires. Unventilated salamanders and certain operations such as blast furnaces in steel works will produce carbon monoxide. Carbon monoxide is a chemical asphyxiant that combines with hemoglobin.

There are also what we call simple asphyxiants such as nitrogen, helium and methane. Some of these are fire hazards and some are used as blanketing agents and have caused death by asphyxiation. The critical thing here is the amount of oxygen. If the oxygen level is only six per cent and the rest of the atmosphere is nitrogen, employees will collapse surprisingly fast as if struck on the head. We normally breathe air which is roughly 80 per cent nitrogen. If we were to increase the amount of nitrogen to 94 per cent, we would very quickly collapse from lack of oxygen.

A second group of gases and vapors make their presence known through irritation. These include ammonia, sulphur dioxide, chlorine, hydrochloric acid, phosgene and nitrogen dioxide. In the case of ammonia, sulphur dioxide, chlorine and hydrochloric acid, employees will voluntarily vacate the area when these substances are accidentally released. The reason for this is because they are immediate irritants that are readily soluble in body tissue and hit "high up." They hit the eyes, the nose and the throat. Phosgene and oxides of nitrogen are not as soluble in tissue fluids. Thus, they can be inhaled in serious amounts without too much discomfort at the time of exposure. Later on, after reacting with tissue fluids to produce acids in the lung tissue, a fatal pulmonary edema may result.

Systemic poison

An example of a systemic poison is lead which accumulates in the body and is stored there. It has various effects once it is stored. It can cause gastro-intestinal effects. What we see mostly today is constipation and colic. Most of the 10 to 20 cases a year in Ontario which end up as compensable tend to be the gastro-intestinal type with constipation and colic as well as an elevated blood lead level. We very rarely get neurological changes characterized by paralysis or brain changes.

The controversial area now is the effect of lead on reproduction. There are some who feel that blood lead levels in the circulating blood of a pregnant woman can be such that they do not harm the mother but may affect the developing fetus. This presents a real dilemma in terms of human rights because women can say they are being discriminated against if they are prevented from working with lead during the fertile period of their life. However, the fact is a fetus is more sensitive than a female or male adult. There is also some concern that lead may cause reproductive effects in males and this is under study.

Summary

I have described how environmental agents can enter the body and cause harmful effects, some of the defense mechanisms of the body, and some of the common health hazards in the workplace. The key point for all of us is to review our workplaces, try to recognize and assess the health hazards which may exist, and then take action to control these hazards.

Reprinted with permission from Accident Prevention, based on a speech presented at Occupational Health in the Workplace program held in Toronto sponsored by the Canadian Occupational Health Association and Industrial Accident Prevention Association.

If the oxygen level is only six per cent and the rest of the atmosphere is nitrogen, employees will collapse surprisingly fast...

A case history

Training for toxic materials handling

by James E. Gillan

ABSTRACT. *Every new OSHA health standard will require training in order that workers will become knowledgeable about toxic hazards. The following case history illustrates how one group of line managers provided adequate training to accomplish the company's goal of increased safety.*

As everyone familiar with the field knows, in the coming decade OSHA will issue increasingly rigid occupational health standards which will affect both businesses and employees. That they will is both inevitable and desirable. Each standard will require employers to train and retrain workers to make them knowledgeable about the occupational health hazards of specific toxic materials. The training will have to be complete and forthright.

Providing adequate training will be a challenge to industrial managers and more particularly to safety professionals. How will it be accomplished for instance, within smaller companies where line managers do not have ready access to staff safety professionals? At Gould Inc., one group of line managers met this challenge with commendable results.

To properly understand the case history it is necessary to have some knowledge of the organization of the corporation it involves.

We are a large corporation with annual sales in excess of $1.6 billion, consisting of more than 40 operating divisions, each of which manufactures and sells a line of electrical, electronic or industrial products to a well defined market. The corporation rigidly controls each operating division by establishing financial goals, by requiring and reviewing annual plans and long-range business strategies, and by monitoring capital expenditures. Excepting these and a few other normal business constraints, the divisions are granted considerable autonomy however, and, in fact, are operated as individual business enterprises. A small operating division has a great deal in common with a small independent company.

Safety organization is consistent with the corporation's philosophy. Safety is by policy and practice a line management responsibility. Line managers are expected and required to consider safety as one of their primary personal responsibilities. They are encouraged to exercise initiative in resolving the safety problems unique to their facilities and businesses, and their performance is continuously monitored by the Corporate Safety Department.

The Metals Division can be considered a small business that is almost totally involved in processing a single and classic toxic material —

Working Safely With Lead

GOULD INC., METALS DIVISION EMPLOYEE HANDBOOK

GOULD $2.00

inorganic lead. It faces many of the training problems mentioned at the outset. It operates three secondary lead smelters and three lead oxide plants at five widely scattered locations. The business is capital intensive, so total employment is only about 260 employees including a modest divisional staff with about 15 managers and professionals.

For several years the Metals Division had aggressively upgraded its manufacturing buildings and equipment to reduce lead dust and fume exposures to its workers and the surrounding communities.

Millions of dollars were invested in ventilation and pollution control equipment. Over the past two years more than $500,000 was invested in nonproductive amenities such as shower rooms, locker rooms and cafeterias to support the need for excellent personal hygiene practices by its lead workers. Environmental monitoring was conducted monthly to determine the lead-in-air exposure. All ventilation equipment was inspected daily and evaluated quantitatively monthly. A rigorous program of medical surveillance was adopted involving complete medical examinations and frequent blood sampling to continuously monitor the level of lead absorption of each lead worker.

All of the above activities are relatively standard practices within the lead industry. They are essential for an adequate health maintenance program for lead workers. However, management still found it extremely difficult to consistently comply with existing exposure limits, limits which will soon be lowered by the issuance of a new OSHA Inorganic Lead Health Standard. Much progress has been made in lowering lead-in-air and blood lead readings; but futher improvement was becoming extremely difficult at the same time it was becoming necessary. Something beyond the Division's previous commitment was needed to comply with toughening standards.

It was obvious that all production workers needed to become involved in the company's program if there was to be further progress. Work habits and housekeeping practices are important to limiting the generating of lead dust. Personal hygiene and the proper use and care of personal protective equipment are vital to limiting the inhalation and ingestion of lead. These problems had to be attacked jointly by management and labor in an across-the-board commitment to a total environmental safety program.

Many of the employees had a general understanding of the toxic qualities of lead and the necessary precautions. However, there has always been some reluctance within the lead industry to provide detailed and specific training on the symptomatology of lead intoxication. Most of the early symptoms of lead intoxication are also symptoms of very common illnesses which are not necessarily work-related. The diagnosis of plumbism (lead poisoning),

Employees of Gould's Metal Division discussing their "Working Safely with Lead" booklet and program, left to right: Neal Wentz, Personnel Manager; Dan Verser, Manager of Operations; and Carole Johnson, Executive Secretary to the General Manager.

as opposed to an elevated level of lead absorption, is a difficult diagnosis to make, and is frequently based on an allegation of subjective clinical symptoms. This situation can obviously be troublesome.

Recognizing the need for a more extensive lead health training program, decision was made to "go all out" on a program of worker education and training. Every employee was to be taught as much as possible about the toxicology and proper handling of lead.

The first step in the overall program was to obtain an instructional pamphlet to serve as a basis for uniform instruction and future reference.

Lead is one of the oldest, most commonly used toxic materials known to man. There is extensive literature on the subject and soon much of this was quickly gathered. The materials were written for a wide range of readerships and consequently the language used involved various levels of difficulty and many specialized vocabularies. There was no literature which related to Gould's specific situation. It was quickly determined the Division needed its *own* pamphlet if it hoped to effectively communicate the required information as it related to them — in plain, understandable English.

Knowing what to include was not a major problem. The final text actually parallels closely the types of information the forthcoming OSHA standard will require for employee training. Much time was spent on honing the existing language used in other literature so the pamphlet would not talk over the head of the workforce, or talk down to them. It had to communicate honestly and effectively.

All management personnel within the Division, most importantly, every plant manager became involved in drafting and rewriting. The project was completed in just over six months. But the time spent was worthwhile, because all managers in the Division felt they had made a contribution to the project. They were each thoroughly familiar with the text and were also committed to its success — an important aspect of any safety program. The technique of multiple authorship is important; we recommend it to others. It is questionable if the total program would have been as well accepted if the training material was developed by a staff professional working in isolation.

Illustrations were needed to lighten the text and introduce a touch of humor. Photographs were taken at the plants. These were later supplied to illustrators so artists could properly characterize lead workers and their work environment.

At this point the entire project was turned over to an advertising agency who frequently worked with the Division's Marketing Department. In consultation they established the format, developed illustrations and provided the professional touch that is important but difficult for the layman.

One of the Division's plants has an extremely high percentage of Spanish-American employees so a Spanish language version of the text was also decided upon. The translation was made by a graduate student at a local university at modest cost.

Finally, the pamphlet was contracted to a local printing company. By using in-house talent and discreetly going outside to assure professionalism when needed, the Metals Division was able to obtain a generous supply of their own totally tailor-made instruction pamphlet, "Working Safely with Lead". Out-of-pocket costs came to a few thousand dollars — most reasonable considering the importance of the publication.

Any small company can obtain the same results by using a similar approach. In many cases, companies will have to resort to developing their own personalized instructional material if it is to be targeted to their own requirements and really "do the job".

Developing potentially effective materials is a good start, but it is hardly worth the effort if management merely passes out pamphlets and considers the job completed. How did the Metals Division utilize their material? The program at their plant in Omaha, Nebraska was typical.

The Omaha plant conducted a series of three, off-site dinner meetings to accommodate their multi-shift workforce. Personnel manager Neal Wentz and the plant manager jointly spent about one hour carefully reviewing and completely explaining the information on every page of the 15-page, 5½" x 8½" pamphlet. The discussion included:

Your Health and Lead
How lead enters the body.
What happens when the body absorbs too much lead?
What are the symptoms of lead intoxication?
How can you be sure that you are not absorbing excessive amounts of lead?
How are excessive lead absorption and lead intoxication treated?

Working Safely with Lead
Personal hygiene.
Safe work habits.
Housekeeping.
Respirator protection.

The presentation was completely forthright. No known relevant information was withheld or its significance diluted. The emphasis was as stated in the pamphlet: "Lead is only a potential hazard, but it can become a real hazard if you ignore it."

Possible audience reaction had been the subject of many debates. Would the employees be antagonistic, disinterested, bored or would they react in other non-productive manners?

Their reaction was the best possible. The presentation was followed by a vigorous, two-hour question and answer period with almost total participation. The employees demonstrated an intense interest in virtually every aspect of the problem. They reacted very positively to the concern demonstrated by management and to the fact that management had obviously taken them completely into their confidence. They were especially interested in learning the underlying reasons for many of the established work practices.

The pamphlet will, of course, be used for training all new employees. It will not merely be handed to the employee as something to read. The supervisor will spend at least one hour going over the material thought-by-thought on the first day of employment. On the rear page the plant manager will personally sign a statement which includes in part: "I pledge to you to do everything within my means to provide a safe and healthful place for you to work." The employee will sign a statement acknowledging he or she has read and discussed the handbook with the supervisor. The supervisor will sign that: "The contents of the booklet have been discussed with the above named employee, and I am personally satisfied that he understands its contents."

Retraining will of course, be necessary. Once is never enough. The Division will schedule a form of refresher training on an annual basis, not only to refresh memories of the hazards of lead; but to provide any new information as it becomes available.

Are there measureable or observable results from this program? Yes, there are. Blood lead and lead-in-air readings have descended and continue to descend to lower levels. Housekeeping and work practices have improved; and work-force morale is noticeably higher according to Division management.

Conditions change constantly at any plant. With so many variables it is always difficult to clearly identify cause and effect. However, managers throughout the Metals Division are convinced that their forthright, prompt and unique efforts to train and inform their lead workers in the toxicology and proper handling of lead have been an unqualified success — one that other companies can also achieve at relatively little expense.

As OSHA regulations mandate improved health standards in coming years, we will find again and again that there is no existing literature or program which relates specifically to our company's situation. Personalized safety training programs will become increasingly, a demand and requirement to be met by individual companies.

James E. Gillan, CSP, is Director, Corporate Safety for Gould Inc., Rolling Meadows, Ill. He has worked in loss prevention for almost 30 years. Prior to joining Gould, Mr. Gillan was manager of corporate safety for ICI America Inc., Wilmington, Del. He holds bachelor degrees in education and electrical engineering, the latter from Rensselaer Polytechnic Institute. He is a Past Chapter President, a Professional Member, and at present a member of the Greater Chicago Chapter.

Toxic effects of wood dust exposure

by Angelo Meola

Wood has been used since prehistoric times and except for fire, the hazards of wood dust have been largely ignored. When health problems developed in woodworkers, they were usually thought to be the result of microorganisms such as bacteria and fungi or a non-specific irritation from the dust particle. Later concerns were the adhesives, solvents and other chemicals used in the workplace, and residual amounts of fungicides and insecticides in the wood. More recent work has been concerned with identifying toxic properties of the wood itself.

Solvents, adhesives, fungicides, insecticides and microorganisms all produce health effects. In addition, wood dust has a large surface area to volume ratio which permits adsorption of workplace chemicals that are later released in the body. However, not all of the health effects can be attributed to these causes. The species-specific reactions in regard to asthma and sino-nasal cancer offer evidence that some constituents of wood dust are toxic. The use of solvents, adhesives and other chemicals is not species-specific and neither is the presence of microorganisms.

Wood variability

Wood should not be thought of as a single entity. The composition of wood is extremely variable and analysis is usually divided into cell wall constituents and extraneous components (also referred to as extractives). The cell wall constituents resemble a fiber-reinforced plastic.

The polysaccharides such as cellulose form the fiber network with lignin as the binder. The composition of lignin varies from species to species, but it is a mixture of polymers based on a phenylpropane unit. The extraneous components are chiefly made up of tannins, terpenes, volatile oils, alcohols, aldehydes, and organic acids.

Not all of the extraneous components are in all species and the relative proportions differ. Wood from trees of the same species vary in composition depending on soil and climatic conditions. Even different parts of the same tree have a different composition. For example, the roots and lower trunk contain more silica than the upper trunk and crown.[1] Kiln drying, curing agents and preservatives will all affect composition.

A study by Levin and Nilsson demonstrates the effect of preservatives. Analysis of dust samples from a sawmill handling logs that had been dipped in a 2% solution of pentachorophenol showed residues ranging from 100 to 800 parts per million of tetrachlorophenol and from 30 to 400 parts per million of pentachlorophenol. The variance in the residue levels is probably due to the treatment method. Dipping would produce a concentration gradient measured along the radius, and the dust would have different residues depending on what part of the log had been cut.

Undoubtedly, the heat produced in machining and sanding the wood alters the chemistry. The result of this variability is that wood dust cannot be treated as a single agent.

Effects of wood dust exposure

At least fifty varieties of wood are capable of inducing contact dermatitis.[2] The other effects of wood dust exposure identified so far are:

(1) Impairment of nasal clearance
(2) Obstructive respiratory changes
(3) Asthma
(4) Sino-nasal cancer

Even the wood chips used for bedding material have an effect on laboratory animals. The use of eastern red cedar as a bedding material increases the incidence rate of spontaneous tumors in certain mice. C3H-A^{vy} mice (a particular genetic strain) have an incidence rate between 90% and 100% of spontaneous liver and mammary tumors. However, when these mice were bred and maintained in Australia, the rate dropped to 29%. Douglas fir sawdust was used for bedding the animals in Australia as opposed to cedar shavings in the United States. The high incidence rate can be restored if U.S. bedding material is used in Australia.[3]

(1) Impairment of nasal clearance

Normally, nasal mucous is transported at an average rate of six millimeters per minute from the nose to the pharynx and swallowed. Each individual has a characteristic rate ranging from less than one to more than twenty millimeters per minute. Black, et al, in 1974 and Anderson, et al, in 1977 demonstated a decreased flow rate for nasal mucous among woodworkers. The studies used different methods to measure the flow rate. Black measured the time it took to transport a radioactively tagged polystyrene particle and Anderson measured the transit time of a blue dyed saccharine particle.[4,5] The consistent results indicate that the change in flow rate is not a stratagem of the experimental method. The authors also stated that the change is reversible upon removal from wood dust exposure.

Anderson defined mucostasis as a transit time greater than forty minutes from the anterior end of the middle turbinate (air passages in the nose) to the pharynx. He found a direct proportionality between the number of subjects with mucostasis and dust exposure. Exposure to a high concentration (twenty-five milligrams per cubic meter) of an inert plastic dust produced no change in the flow rate. The plastic had the same size distribution as wood dust, therefore, a chemical action rather than physical blocking is probably involved.[5]

(2) Obstructive respiratory changes

Contradictory findings have been reported regarding obstructive respiratory changes. In L. W. Whitehead's article on the pulmonary function status of woodworkers, he reports that exposure to either dry hardwood dust or dry pine dust is related to obstruction respiratory changes as measured by the ratio of the forced expiratory volume one second to the forced vital capacity (fev_1/fvc). This exposure also significantly lowers values of the maximal mid-expiratory flow rate (MMEFR) compared to theoretical values.[6]

However, Anderson found no significant difference in fev_1 or MMEFR compared to theoretical values and also no difference between subjects with high (more than five milligrams per cubic meter) versus low (less than five milligrams per cubic meter) dust exposure.[5]

Instead of concentration as a measurement of exposure, Whitehead had used a dust exposure index, which he defined as the product of the average dust level in a department and the number of years the employee worked in that department.[6] This type of index would be more likely to identify any effects that depend on cumulative exposure.

L. Michaels performed an autopsy on two wood workers and reported that the lungs exhibited emphysema and fibrosis and contained clusters of foreign body giant cells. Also found were roughly spherical dust particles, ten to fifteen micrometers in size in the lungs.[7] No explanation was offered in the article as to why particles of this size reached the lungs. Normally, particles greater than five micrometers in size are removed from the air stream by impaction in the nose.[8]

Consistent results were reported in animal experiments by Bhattacharjee. Dust suspensions of sheesham and mango woods were intra-tracheally injected into the lungs of guinea pigs. Animals were sacrificed at sixty and ninety days and the lungs examined. Dust laden macrophages, foreign body giant cells, grade 2 fibrosis and emphysema were found in the animals exposed to mango dust. In the animals exposed to sheesham, Bhattacharjee found a diffuse reaction, foreign body giant cells, emphysema and grade 1 fibrosis. There was a disintegration of the giant cells by ninety days.[9]

(3) Asthma

A hypersensitivity reaction leading to asthma has been reported from exposure to a number of woods including the commonly used woods: western red cedar, cedar of Lebanon, oak, mahogany, and redwood.[10-20] There are two clinical patterns. The first is characterized by immediate onset of symptoms and rapid reversibility. The second is a delayed onset (five to eight hours) and more gradual reversibility. Some woods can produce both types of reaction.[20]

Connecting the asthma to wood dust exposure is difficult because frequently the subject has worked with wood for years with no reaction. He normally has no history of bronchial asthma and a skin test with wood dust extract is usually negative. The sensitization develops on a relatively short exposure to the offending wood. It begins as eye and nose irritation and then is followed by an ir-

ritating nonproductive cough in late afternoon or at night. A single exposure, in a sensitized individual, can cause symptoms for two or three consecutive nights which clear up during the day. The only effective diagnostic test is a challenge with the offending wood dust.[18]

Plicatic acid has been identified as the agent in western red cedar; but this compound has not been isolated in any other species. Type and severity of reaction to plicatic acid seems to depend on the degree of sensitization, rather than dose. Repeated exposure to the same dose causes the late reactions to occur earlier and be more severe and prolonged.[13]

The asthmatic reaction is species-specific. In one case study, two subjects who exhibited asthma, one on exposure to iroko (a tree found in West Africa) and the other on exposure to western red cedar, were challenged with the opposite wood. No reaction occurred.[19] In another case study, six employees in one factory had a reaction to cedar of Lebanon dust. Another employee in the same factory had no reaction even though he had a history of asthma when exposed to western red cedar.[16]

(4) Sino-nasal cancer

Adenocarcinoma of the nose and sinus is extremely rare with an annual incidence of 0.9 per million in the total population. Two-thirds of these are woodworkers.[5] The association has been shown with woodworking crafts within the furniture industry and not with woodworking, in general. There is also no association with other workers in the furniture industry such as finishers, polishers and upholsterers.[21]

The machining and sanding operations in a furniture factory are usually physically separated from the other operations because the dust would ruin the finish on the furniture. This tends to isolate the workers exposed to wood dust from the other workers.

Studies indicating an association between wood dust exposure and adenocarcinoma of the nose have been done in several countries including Great Britain, Italy, the United States, Sweden, and Denmark.[5,21-26] The agent appears to be related to species of wood. Milham found no excess of nasal cancer in a study of industries which worked, primarily, with soft woods (pine, spruce).[27] The furniture industry worked primarily with hard woods (oak, beech and maple).[21]

Latency periods have been reported as low as nine years and as high as seventy years. The case with the nine year latency period worked exclusively with oak.[26] Varying dose levels and, perhaps, the cumulative effect of exposure to a particular species as opposed to exposure to all wood dust is involved. Epidemiological studies involving wood dust are difficult because there is no way of ascertaining either the specific identity or the dose level of the wood dust involved. Woodworkers normally work with a variety of woods and, although the genus and species, method of drying, and any chemical treatment should be identified, it is often not possible.

Controlling occupational exposure

There are varying opinions on what constitutes a safe level of exposure to wood dust. In the United States, the OSHA standard for exposure to airborne wood dust is five milligrams per cubic meter. In the Soviet Union, the standard is four milligrams per cubic meter for dusts that contain less than 10% silica and two milligrams per cubic meter for dusts with over 10% silica.[28] Neither of these standards takes into account the range of toxic effects of different woods.

Several attempts have been made to rate the relative toxicity of woods. Kadlec and Hanslian, in 1972, suggested a numerical system.[28] It relied heavily on the allergenic and dermatitis reactions to make the divisions.

To develop in vitro tests for predicting the toxic effects of different woods, Bhattacharjee examined the hemolytic activity and macrophage cytotoxicity of sheesham and mango. These woods were chosen because they are commonly used in the furniture industry in India where the study was done. He concluded that hemolytic activity may be an index of acute toxicity, and microphage cytotoxicity may be an index of fibrogenicity.

The impairment of the clearance of nasal mucous may be the most useful index. The effect on health of the slower clearance rate is not known, but it increases the retention time of the wood dust (and any adsorbed material) in the nose. Exposure limits for individual wood species could be

set at a level below that which would cause impairment.

Establishing permissible exposure limits is the job of a toxicologist, but even the current 5 microgram per cubic meter limit is not followed in many cases. It is not uncommon to find wood dust levels 5 to 10 times as high. Frequently, both owners and employees in woodworking shops think of wood dust as an inert nuisance dust. The safety professional is faced with the task of informing them that a health hazard exists.

References

1. Browning, B. L., Editor. *The Chemistry of Wood;* New York, Interscience Publishers, 1963.
2. McCord, Carey P. "The Toxic Properties of Some Timber Woods." *Industrial Medicine and Surgery;* 27, pp 202-204, 1958.
3. Sabine, J. R. et al. "Spontaneous Tumors in C3H-A^{vy}FB Mice: High Incidence in the United States and Low Incidence in Australia." *Joural of the National Cancer Institute;* 50, 5, pp 1237-1242, 1973.
4. Black, A. et al. "Impairment of Nasal Mucociliary Clearance in Woodworkers in the Furniture Industry." *British Journal of Industrial Medicine;* 31, pp 10-17, 1974.
5. Anderson, H. C. et al. "Nasal Cancers, Symptoms and Upper Airway Function in Woodworkers." *British Medical Journal;* 34, 3, pp 201-207, 1977.
6. Whitehead, L. W. et al. "Pulmonary Function Status of Workers Exposed to Hardwood or Pine Dust." *American Industrial Hygiene Association Journal;* 42, 3, pp 178-186, 1981.
7. Michaels, L. "Lung Changes in Woodworkers." *Canadian Medical Association Journal;* 96, pp 1150-1155, 1967.
8. Proctor, Donald F. "The Upper Airways 1. Nasal Physiology and Defense of the Lungs." *American Review of Respiratory Disease;* 115, pp 97-129, 1977.
9. Bhattacharjee, J. Wynne. "Wood Dust Toxicity: In Vivo and In Vitro Studies." *Environmental Research;* 20, 2, pp 455-464, 1979.
10. Raghuprasad, P. K. et al. "Quillaja Bark (Soap Bark) Induced Asthma." *Journal of Allergy and Clinical Immunology;* 65, 4, pp 285-287, 1980.
11. Bush, Robert K. et al. "Asthma Due to African Zebrawood (Microberlinia) Dust." *American Review of Respiratory Disease;* 117, pp 601-603, 1978.
12. Chan-Yeung, M. and Aboud, R. "Occupational Asthma Due to California Redwood (Sequoia Sempervirens) Dusts." op. cit.; 114, pp 1027-1031, 1976.
13. Chan-Yeung, Moira et al. "Occupational Asthma and Rhinitis Due to Western Red Cedar (Thuja Plicata)." *American Review of Respiratory Disease;* 108, pp 1094-1102, 1973.
14. Dopico, G. A. "Asthma Due to Dust From Redwood (Sequoia Sempervirens)." *Chest;* 73, 3 pp 424-425, 1978.
15. Eaton, K. K. "Respiratory Allergy to Exotic Wood Dust." *Clinical Allergy;* 3, pp 307-310, 1973.
16. Greenberg, M. "Respiratory Symptoms Following Brief Exposure to Cedar of Lebanon (Cedra Libani) Dust." *Clinical Allergy;* 2, pp 219-224, 1972.
17. Howie, A. D. et al. "Pulmonary Hypersensitivity to Ramin (Gonystylus Bancanus)." *Thorax;* 31, pp 585-587, 1976.
18. Milne, James and Gandevia, Bryan. "Occupational Asthma and Rhinitis Due to Western (Canadian) Red Cedar (Thuja Plicata)." *Medical Journal of Australia;* 2, pp 741-744, 1969.
19. Pickering, C.A.C. et al. "Asthma Due to Inhaled Wood Dusts—Western Red Cedar and Iroko." *Clinical Allergy;* 2, pp 213-218, 1972.
20. Sosman, Abe J. et al. "Hypersensitivity to Wood Dust." *The New England Journal of Medicine;* 281, 18, pp 977-980, 1968.
21. Acheson, E. D. et al. "Nasal Cancer in Woodworkers in the Furniture Industry." *British Medical Journal;* 2, pp 587-596, 1968.
22. Acheson, E. D. et al. "Adenocarcinoma of the Nasal Cavity and Sinuses in England and Wales." *British Journal of Industrial Medicine;* 29, pp 21-30, 1972.
23. Cecchi, F. et al. "Adenocarcinoma of the Nose and Paranasal Sinuses in Shoemakers and Woodworkers in the Province of Florence, Italy (1963-1977)." op. cit.; 37, 3, pp 222-225, 1980.
24. Roush, G. C. et al. "Sinonasal Cancer and Occupation: A Case-Control Study." *American Journal of Epidemiology;* 111, 2 pp 183-193, 1980.
25. Engzell, U. "Nasal Cancer Associated with Occupational Exposure to Organic Dust." *ACTA Otolaryngol;* 86, 5-6, pp 437-442, 1978.
26. Engzell, U. "Occupational Etiology and Nasal Cancer, an Inter-Nordic Project." *ACTA Otolaryngol (Suppl);* 360, pp 126-128, 1979.
27. Milham, Samuel. "Neoplasia in the Wood and Pulp Industry." *Annals of the New York Academy of Sciences;* 271, pp 294-300, 1976.
28. Radlec, K. and Hanslian, L., "Wood." *Encyclopedia of Occupational Health and Safety;* New York, ILO, McGraw Hill, 1972.

The clampdown on electrical hazards

by Milton Leonard,
Senior Editor,
Machine Design Magazine,
Cleveland, Ohio

Electric power is our most useful servant. Used carelessly, however, it imposes harsh penalties in loss of life, injury, and property damage. A growing concern for safety at home and on the job is bringing tougher standards for all electrical equipment.

Electric shock kills about 1,000 Americans each year and injures thousands more. Fire damage from faulty electrical equipment totals millions of dollars annually. These alarming statistics have brought a closer working relationship between governmental and industrial agencies in establishing tougher enforcement of more severe safety standards.

The Occupational Safety and Health Administration was formed over three years ago to protect workers in industrial plants and on construction sites. OSHA's requirements have been primarily the concern of plant owners, not manufacturers of equipment sold to the plant. Two years ago the Consumer Product Safety Commission was established to formulate and enforce safety standards for the design and manufacture of consumer products used in the home, school, and for recreation.

By applying their enforcement powers to the applicable standards of private and government standards makers, like NEMA, UL, NEC, and ANSI, these two agencies have become vigilant watchdogs of safety. Until recently, OSHA and CPSC jurisdictions came close to overlapping in some areas and left gaps in others, so that many electrically-powered products still have weak or nonexistent safety controls. These gaps are being gradually closed, however, and there is little reason to doubt that eventually all electrically powered products will be subject to stringent safety standards.

Current is the culprit

A voltage difference across the human body is necessary for shock. But the resultant current flow does the damage. Human-body response to alternating current flow can be classified into three levels.

Perception or reaction current is that level of al-

OSHA Means Business

Last year OSHA violations resulted in about $10½ million in penalties. Subsequent corrective action cost violators approximately $11 million more. Electrical violations of the OSHA-enforced NEC numbered about 11,000.

- 50% of electrical violations related directly to 1971 NEC Section 250-45 which regulates grounding of equipment connected by cord and plug.
- 12% related to Section 400-4 which outlines conditions under which flexible cords and cables are prohibited.
- 12% related to Section 250-42 which governs grounding of permanently mounted equipment, such as machine tools.
- 9% related to Section 400-5 which regulates flexible splices and connectors in flexible power cords and cables to equipment.
- 4% related to Section 250-51 which regulates general grounding procedures.
- 3% related to Section 400-10, the standard for tension on power wire terminals.
- 2½% related to Section 250-5 which regulates the use of connectors for ac system grounding.

Right now the manufacturer who uses electrical factory equipment is shouldering most of the responsibility for OSHA-enforced electrical safety. The designer of this equipment has only had to provide proper equipment interface with OSHA-regulated, in-plant power facilities. Soon, however, factory-equipment designers may be more responsible for their product's safety, as already are designers of consumer products which are regulated by the Consumer Product Safety Commission.

---

*Reprinted from MACHINE DESIGN, January 9, 1975. Copyright, 1975, by The Penton Publishing Company, Cleveland, Ohio.

ternating current that produces a slight tingling sensation. The startling effect from this sensation could produce involuntary muscular reaction that might cause injury. Laboratory tests show that the lowest perceivable current at 60 Hz varies in different individuals. Less than 1% can sense current levels as low as 0.3 mA. However, the mean value of perception current in men is about 1.1 mA. The mean perception level for women is about 0.7 mA. These levels do not damage human tissue. As a result, perception current has been established at 0.5 mA.

Let-go current is the maximum current at which an individual grasping a conductor can release it by using the muscles directly affected by the current. Let-go current, averaging about 16 mA for men and 10.5 mA for women, also does not damage human tissue.

Lethal currents begin at only slightly higher levels. When a current in excess of 18 mA flows through the chest cavity, the chest muscles contract to stop

The Human Body as a Circuit Element

When the human body becomes part of an electrical circuit, it follows Ohm's law. Current flow through the body is $I = E/Z$, where Z is total circuit impedance. The amount of current flow through the body, and thus severity of shock, depends on several factors.

Body Resistance: At 60 Hz, the human body essentially acts as a resistor. Minimum electrical resistance of the body between major extremities is generally accepted as 500Ω. However, surface or contact resistance—the chief initial current-limiting factor—varies between different parts of the body and between individuals. Dry skin resistance varies from 100 to 300 $k\Omega/cm^2$. But wet skin has only 1% the resistance of dry skin. Furthermore, if shock current flows for a few seconds, blisters develop which further decreases surface resistance.

Voltage Amplitude: The voltage necessary to produce shock current depends on contact resistance and total circuit impedance. Commercial 120 Vac voltage is lethal when contact resistance is low. At high voltage contact resistance is not a factor, since voltages in excess of 240 Vac punctures the skin upon contact. Under this condition only internal body resistance limits shock current.

Frequency: At 60 Hz body impedance is essentially resistive. As frequency increases, body impedance begins to act as a resistor-capacitor network and becomes nonlinear. At about 50 kHz body impedance may decrease more than 50%, and contact resistance is negligible.

Voltage frequency also affects perception, let-go, and fibrillation currents. A 7-mA current is required at 5 kHz to produce tingling. In the region of 100-200 kHz perception changes from tingling to heat. At 5 kHz let-go current increases threefold over the 60-Hz value. The current required to cause fibrillation also increases at higher frequencies.

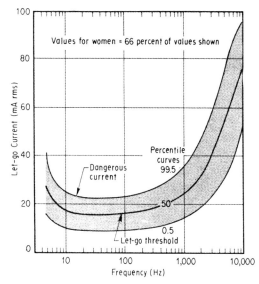

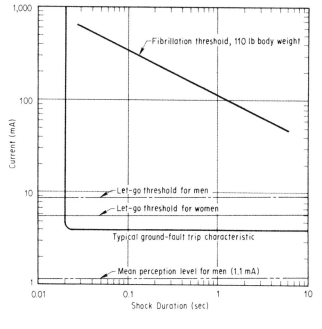

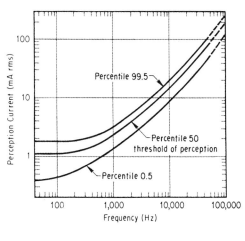

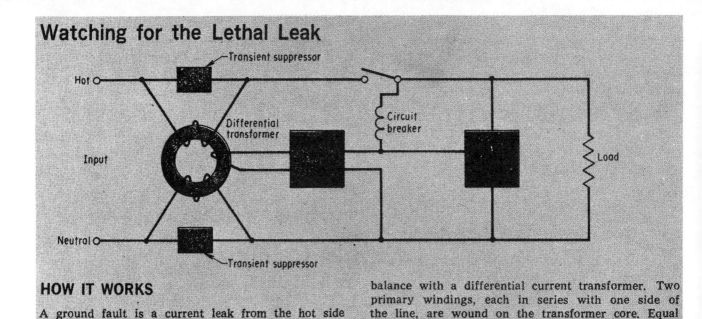

Watching for the Lethal Leak

HOW IT WORKS

A ground fault is a current leak from the hot side of the line through a path that bypasses the load to ground. This current leak causes an imbalance between the hot and neutral wires to the load. A typical ground-fault interrupter senses the imbalance with a differential current transformer. Two primary windings, each in series with one side of the line, are wound on the transformer core. Equal current in both lines cancel each other to produce no transformer signal in a third sensor winding. However, current imbalance generates a sensor output which is amplified to trip a circuit-opening device.

---

breathing. Maintained current flow results in loss of consciousness and, eventually, death. *Ventricular fibrillation* is another potentially lethal result of shock in which the heart ceases its rhythmical pumping action, and instead feebly quivers to effectively stop blood circulation. The heart rarely recovers spontaneously from this condition.

Extrapolations from data compiled from experiments on animals show that current levels necessary for producing fibrillation in humans depend on shock duration, body weight, current-flow path, flow duration, and current magnitude. It is generally believed that the heart of a normal adult is likely to fibrillate when shock current in milliamperes exceeds $116/t^{1/2}$, where t is shock duration in seconds.

Shock current that greatly exceeds the level necessary to produce fibrillation may completely stop heart action, seriously burn body tissues, damage the nervous system, and stop breathing—all potentially lethal conditions.

Traditional methods for protecting users of electrically powered products against shock include double insulation (insulated interior parts plus housings of dielectric material), grounding, and the use of isolation transformers. But, electrical accidents still occur.

Double insulation, for example, does not protect against defects in the cord, plug, and receptacle.

Grounding a metallic housing of a product helps when interior voltage-carrying components inadvertently contact the housing. But unless the ground path has zero impedance, human contact with the housing can still result in shock.

Isolation-transformer effectiveness is easily nullified by faulty wiring and worn insulation.

Circuit breakers and fuses protect electrical equipment from overcurrent. They do not prevent shock. For example, fuses are available with ratings down to 2-mA, but they are not designed to protect against ground faults.

New safety techniques

A significant number of electrical accidents have involved the power-cord/receptacle interface of electrical equipment. The two-prong power plug with a third grounding prong undoubtedly eliminates many accidents. For increased safety, 3-prong power receptacles are available with shutters across the power-blade slots that do not open unless the plug has a grounding prong. Thus, a live power connection cannot be made until a ground connection is completed. Additionally, encapsulated plastic plugs are replacing older types that had removable fiber discs covering the front of the plug. This transition is a result of the NEC requirement that the front cover for wire terminals in plugs be mechanically secured or be an integral part of the plug. The latter configuration is called *deadfront construction*.

Power connectors are also available with non-metallic cord grippers to prevent strain on the electrical connection when the power cord is pulled. If the nonmetallic gripper contacts a live wire in the cord, it will not present a shock hazard.

NEMA has also been active in this area. To eliminate the hazard of mating electrically incompatible connectors, NEMA—along with wiring-device manufacturers—has developed standard configurations for all ratings of straight-blade and locking power connectors. For example, a 3-pole, 3-wire locking connector with a 15-A, 125-V rating cannot fit a similar device with a 20-A, 250-V rating.

Ground fault: the accident waiting to happen

The new power connectors reduce the very obvious possibilities of electrical accidents. The more subtle and insidious danger, however, is the ground fault. A ground fault results when a current-carrying part of a circuit accidentally contacts any grounded conducting material. The resistance of the path to ground may be high or low, depending on the nature of the contact.

Low-resistance faults draw heavy currents which can trip circuit breakers and open fuses, thereby preventing equipment damage and fire. But when the ground fault has relatively high resistance, such as that of the human body, fault current is not large enough to open circuit breakers and fuses. This same current, however, is sufficient to kill. Furthermore, if the unintentional contact from a live wire to a high-resistance ground is intermittent (such as through vibration) the resultant sparking could trigger a fire—still without opening overcurrent protectors.

Although ground faults have caused accidents and fatalities since the early 1900s, it took the increasing popularity of home swimming pools to bring the problem to surface. The combination of water, submerged lighting, and various electric appliances at poolside was a perfect environment for electrical accidents. Increasing reports of shocks and electrocutions caused the NEC to become concerned.

The search for ground-fault protection culminated in the invention of several devices at home and abroad. The first marketable device in this country, however, was the Ground-Fault Interrupter (GFI) of Charles F. Dalziel, a professor at the University of California.

The operation of a GFI is simple. The device, connected between the power source and the load, compares the current in the two conductors supplying power to the load. Any current leaking from the load circuit to ground creates a current difference between the two conductors. When this leakage exceeds a preset value for a predetermined time, the GFI opens the circuit.

The most widely used type of GFI has a donut-shaped differential transformer through which pass the circuit conductors. Minute current differences are detected by an electronic circuit which then opens a circuit breaker. Thus, a GFI continuously compares the current into and out of the load.

Whenever a leakage current attempts to seek ground through anything but the circuit conductors, the resultant current imbalance causes the GFI to open the circuit. GFIs operate on both two-wire and three-wire equipment.

GFIs were originally developed to protect people in and around swimming pools. Then applications extended to wherever water and electricity are used together or where users of electrical equipment cannot avoid being grounded, such as inside metal tanks or on steel scaffolding. Portable models are becoming a standard tool-crib item for workers using electrical hand tools and outdoor receptacles.

The latest generation of GFIs reflect the advantages of solid-state electronics. They are small enough to be combined with ordinary circuit breakers or duplex receptacles, can detect smaller fault currents, and have faster tripping time for circuit interruption.

GFIs can detect ground-fault current that is only 0.0001 of the load current. They can also discriminate against short-duration voltage transients of up to 8,000 V and retain calibration in temperatures from $-35°$ C to $66°$ C. However, power lines that are about 250 ft or more in length may produce nuisance tripping due to capacitive and resistive leakage current through the insulation.

Successful development and application of GFIs have resulted in the establishment of three important standards: National Electrical Manufacturers' Association Standard for Ground Fault Interrupters; Underwriters' Laboratories, Inc. Standard UL943; and the Canadian Standards Association Standard C22.2, No. 144.

There are approximately two dozen different models of GFIs available. These are classified by *group* and *class*. "Group" designation indicates how fault current is sensed. Group I, the only such designation at this time, indicates GFIs that sense the vectorial sum of currents in power conductors. "Class" designation indicates fault-current trip level. Presently there are two class designations—A and B. A Class A GFI must trip with a ground-fault current of at least 5 mA. It also must trip in $(20/I)^{1.43}$ seconds, where I is fault current in milliamperes. Class B GFIs, for use only in swimming-pool underwater lighting circuits, must trip at 20 mA. Trip time is $(80/I)^{1.43}$ sec, but the trip time-current curve is the same as Class A devices as far as the effect on a person in the water. (Current through a swimmer's body due to a damaged lighting fixture will not exceed 25% of total fault current if the fixture complies with UL Standard 676.)

GFIs are constructed in a variety of configurations. Units for permanent outdoor installation are relatively large and have weatherproof enclosures. Portable indoor models are smaller and are connected to power receptacles through a plug and power cord. Some indoor models may be plugged directly into power receptacles.

For OEM applications, several new models have appeared. One modular design has the GFI circuit components mounted on a 3x4-in. printed-circuit board for mounting in 120-V, 15-A, 60-Hz equipment. Terminals are provided for line, test-switch, and re-set-switch connections. Double-pole interruption protects against line reversal or an open neutral in the line circuit.

More recently, GFIs have been combined with thermal-magnetic circuit breakers and encapsulated in the breaker housing. A test button is provided to simulate a ground fault when pushed for checking both mechanical and electrical operation. Models are available to replace existing circuit breakers with up to 30-A ratings and up to 10-kA interrupting capacity. Thus, these devices protect against both overload and ground faults.

GFIs do not protect against shock caused by line-to-line contact and, with the exception of the combined GFI-circuit breaker, are not intended to replace fuses and circuit breakers. They are intended to complement other safety devices for an added measure of human safety and are effective only when the overall electrical circuit design reflects good safety practices.

Acknowledgement

The author and his publication acknowledge with appreciation the cooperation of these companies in the preparation of this article.

Bryant Div. of Westinghouse Electric
    Corp. . . . . Bridgeport, Conn.

Harvey Hubbell Inc. . . . . Bridgeport, Conn.

ITE Imperial Corp. . . . . Spring House, Pa.

–PS–

Exploiting technology for electrical safety

ABSTRACT. Adherence to the National Electrical Code (NEC) is generally required by federal and local laws in our country. This is looked on as sound, inasmuch as the NEC is constantly being updated to reflect the best current practice. The author discusses the NEC and how it can be utilized by the safety professional with responsibilities for electrical safety.

by Alan Krigman, President,
ICON/Information Concepts, Inc.,
Philadelphia, Pennsylvania

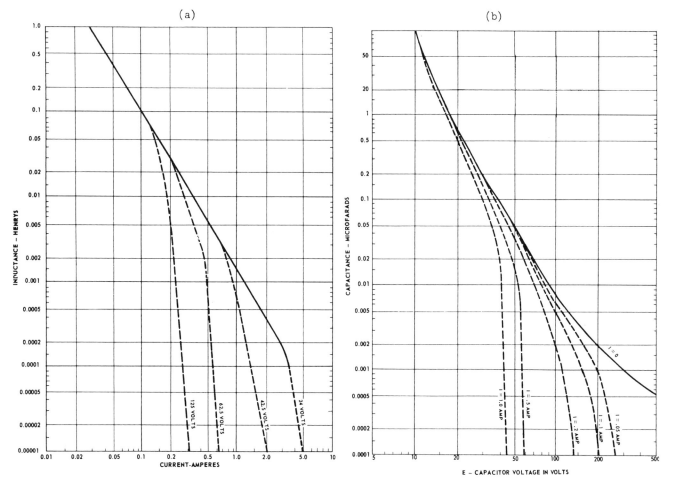

Figure 1

Breakflash calibration curves presented in ISA RP 12.2 and NFPA Standard 493, for the most easily-ignited mixtures of hydrogen in air. Curves represent actual ignition limits; in practice a safety factor of four must be applied to abnormal conditions, depending on the system contact configuration. Circuits may be operated beyond these limits if shown to be safe by breakflash testing. a) series inductance which will not result in ignition of hydrogen atmosphere, for indicated current and voltage. b) shunt capacitance which will not result in ignition of hydrogen atmosphere, for indicated voltage and current.

Laws, codes, and standards concerning electrical apparatus for hazardous environments are intended to guard the public from incompetent or careless engineering. The protection is afforded by adherence to rules and procedures, which reflect opinions of recognized competent practitioners. The documents accordingly govern design and application, and also serve as inspection guidelines.

OSHA and the National Electrical Code

In the United States, federal and local laws generally require adherence to the National Electrical Code (NEC). This is accepted as sound policy, because the NEC is continually updated and revised; and it is widely acknowledged as reflecting the best current practice.

Some recent questions have arisen due to passage of the Williams-Steiger Occupational Safety and Health Act, relating to the use of electrical equipment in hazardous locations. Subpart S of the Act incorporates the NEC as law. However, within the NEC approval of an installation is defined in terms of acceptability to an appropriate enforcement authority, while in the Job Safety Act approval responsibility is vested specifically with the Occupational Safety and Health Administration (OSHA). Unfortunately, provisions of Subpart S covering acceptability are somewhat subjective. In particular, acceptability criteria include listing or approval by recognized testing laboratories, inspection by various authorities of equipment or installations not normally tested by such laboratories, or manufacturer approval of custom-made equipment or related installations. Since electrical safety in industrial plants usually depends on entire connected systems, which are inherently unique to each installation if only by virtue of cable runs, the third acceptability should apply. However, many enforcement offices insist on Underwriters' Laboratory listing as the sole condition of approval. Although this is unnecessary in some cases and meaningless in others, OSHA has declined to issue interpretations, so policies are at best non-uniform and at worst ambiguous.

Intrinsic safety

Articles 500 through 517 of the NEC are concerned with electrical equipment in flammable atmospheres. Detailed rules are given concerning the design and use of **explosion-proof** apparatus for these environments. However, the NEC also states that "equipment and associated wiring approved as intrinsically safe may be installed in any hazardous location for which it is approved, and the provisions of Articles 500 through 517 need not apply." Intrinsic safety is, therefore, viewed as an inherently different concept than explosion-proofing and one in which exploitation of technology rather than adherence to procedure provides protection.

A number of supportive standards and practices

about the author

Alan Krigman is President of ICON/Information Concepts, Inc., a technical information services firm with primary interest in the area of industrial control and instrumentation. Mr. Krigman is a mechanical engineer with a bachelor's degree from MIT and a master's degree from Clarkson College.

His professional experience includes work as the editor of a journal in the field of instrumentation, as a development engineer at the Battelle Memorial Institute, and as an assistant professor of mechanical engineering at Clarkson.

Mr. Krigman is an active proponent of the safety concept advocated in this article. He has participated in a number of seminars on intrinsic safety sponsored by his company and is in the process of launching a newsletter on the subject.

exist, defining what constitutes intrinsically safe installation, and explaining how to evaluate the conformance of specific apparatus with the definition. A list of such documents is given in the bibliography. The references differ from one another in detail, but are similar in essential content. For example, intrinsically safe circuits are broadly defined as incapable of releasing sufficient energy under normal or abnormal conditions to cause ignition of a specific hazardous atmospheric mixture. Likewise, the sources all point out that individual components may be certified as intrinsically safe for use in hazardous occupancies, but that in evaluating the safety of an installation all interconnected parts must be considered.

Establishing safety

The various standards and recommended practices provide three basic approaches for establishing intrinsic safety. All three methods can be traced directly to empirical determinations of the lower explosive limits of various gaseous mixtures, and to the power which a circuit can release in a spark.

The primary method of establishing suitability of apparatus for use in intrinsically safe systems requires testing in an explosion chamber with a spark-producing or breakflash apparatus. The technique is time-consuming and costly. Moreover, results are statistical since approval depends on **inability** to cause ignition in a test of finite duration.

The other two methods depend on the calibration curves of the breakflash apparatus rather than on individual breakflash tests. The calibration data accepted in the United States are plotted for atmospheres containing different hazardous gases, in the form indicated in Figure 1. The curves show current and voltage levels in capacitive and inductive circuits, which are sufficient to cause ignition in the most unfavorable mixture of the particular gas and air. Devices characterized by points sufficiently below the appropriate curves can be accepted for use in intrinsically safe systems without actual testing.

The breakflash calibration curves can be used with a theoretical circuit analysis, assuming various combinations of fault conditions, to calculate whether the current and voltage can ever reach dangerous levels for given circuit impedances. The difficulty with this approach is the lack of assurance that a real circuit is modeled accurately for analysis, or that the worst-case combinations of fault conditions have in fact been investigated.

More effective use of the breakflash calibration curves

More effective use of the breakflash calibration curves is possible for circuits in which the available voltage and current are limited at the source by certified devices such as a transformer or an intrinsic safety barrier. In this case the circuit **impedance** can be examined to determine whether the capacitance and inductance are sufficiently below the limits for given maximum applied current and voltage. Circuit analysis to determine impedance is possible, but again has the limitation that a model may not be valid.

Measurement is also allowable as a means of determining impedance. However, difficulties arise because impedances of devices such as solenoids or moving-coil galvanometer cannot be ascertained using standard instruments.

The measurement problem can be overcome using a comparison technique to ascertain that the unit being investigated cannot release more energy than can a circuit having the limiting values of reactance. This concept provides a means for verifying the safety of a connected circuit as well as of an individual device, since the energy storage and release characteristics of the entire system in the hazardous area can be easily checked at the point where the wiring enters the danger zone.

Compliance

Codes and standards frequently warn against reliance on rules and procedures as substitutes for sound engineering. The U.S. National Fire Protection Association (NFPA) Standard 493-1969 states, for example, that its object is "to provide information for the design and evaluation of equipment depending for safety on the limitation of energy to be used in . . . hazardous locations . . . . It is not an instruction manual for untrained persons but is intended to promote uniformity of practice among those skilled in the art."

Similarly the Instrument Society of America (ISA) Recommended Practice RP 12.2 "is intended to promote uniformity among specialists. It is intended that it be applied only by those who have carefully studied the subject. It is not intended to be an instruction manual for untrained persons."

Compliance with documents is sometimes considered to be problematic. In particular, difficulties have been encountered in connection with intrinsic safety because the technique is a means of using technology **to obviate the need for procedures** to protect the public. The details of the NEC, for example, are irrelevant for systems which are intrinsically safe, and the various standards are not cookbooks with the recipes some seek.

The positive approach

Component and system designers and users who take advantage of technology can best serve the public interest by adopting a positive rather than the usual defensive stand. For example, in intrinsic safety this involves establishing policy documents for an organization which enumerate the various means of implementing and verifying safety under accepted standards. For design purposes the materials should include appropriate specification limits and an explanation of how relevant properties can be checked. The information can be used as part of the product-support literature for customers and during approvals applications. For application purposes the materials would detail purchasing, installation, maintenance, and system modification practices. Suppliers would be asked to conform with the requirements during bids.

In either case the documents are relatively easy to produce. For example, intrinsic safety must depend either on a protective component or on total integrated system design. For the former, the standard can require identification of the protective component and proof that the connected equipment and wiring conform to the corresponding accepted impedance constraints.

Internally-generated intrinsic safety documents should be submitted to the appropriate approval authorities at the earliest possible stages of system design. The materials should be accompanied by references to sections of the NEC and to the various standards and recommended practices under which safety is established. Submission should be accompanied by a request for specific comments on areas of disagreement. The same document later can be submitted in connection with approvals, along with detailed records of engineering decisions during development, installation, and operation. This will help support the contention that the system in the field does indeed conform to the best practicable means available for ensuring electrical safety.

References

1. *Occupational Safety and Health General Industry Standards and Interpretations—Subpart S*, U.S. Department of Labor, October, 1972.

2. Watt, J. H., *NFPA Handbook of the National Electrical Code*, McGraw-Hill Book Company, 1972.

3. "Intrinsically Safe and Non-incendive Electrical Instruments," *Recommended Practice RP 12.2*, Instrument Society of America, 1965.

4. Krigman, A., "Putting Your Fingers on Intrinsic Safety," *Instruments and Control Systems*, September, 1974.

5. Redding, R. J., *Intrinsic Safety*, McGraw-Hill Book Company, 1971.

6. "Shunt Diode Safety Barriers," *Certification Standard SFA 3004*, British Approval Service for Electrical Equipment in Flammable Atmospheres, 1971.

7. "Intrinsically Safe Process Control Equipment," *Standard No. 493-1969*, National Fire Protection Association, 1969.

8. Krigman, A., Redding, R. J., "Test for Intrinsic Safety," *Hydrocarbon Processing*, October, 1974. —PS—

NORTH CAROLINA
STATE BOARD OF COMMUNITY COLLEGES
LIBRARIES
ASHEVILLE-BUNCOMBE TECHNICAL COMMUNITY COLLEGE

DISCARDED

APR 15 2025

3 3312 00300 0684

T 55 .R43 1985

89-0144

Readings in hazard control
and hazardous materials

ASHEVILLE-BUNCOMBE
TECHNICAL COMMUNITY COLLEGE
LEARNING RESOURCES CENTER
340 VICTORIA ROAD
ASHEVILLE, NC 28801